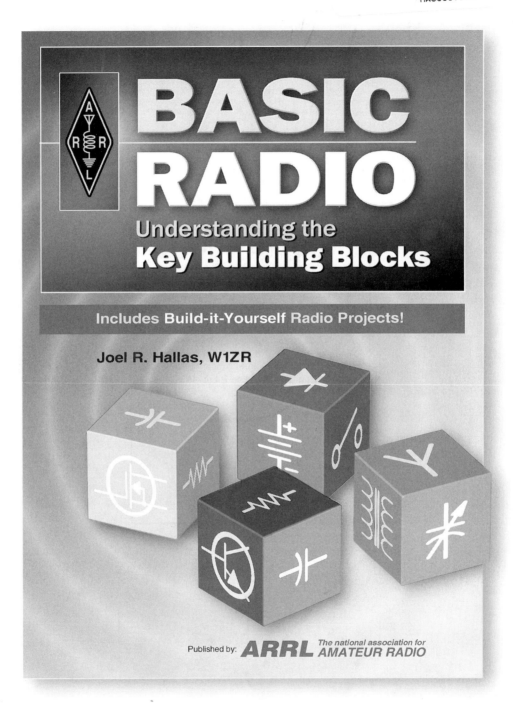

BASIC RADIO
Understanding the Key Building Blocks

Includes Build-it-Yourself Radio Projects!

Joel R. Hallas, W1ZR

Published by: **ARRL** *The national association for* AMATEUR RADIO

Author
Joel Hallas, W1ZR

Production
Michelle Bloom, WB1ENT
Jodi Morin, KA1JPA
David Pingree, N1NAS

Cover
Sue Fagan

MW00837582

Foreword

Radio is all around us. We can't escape it. The radio spectrum follows us wherever we go, whether we use it or not. But we do use it: We listen to the radio in our car, use our cell phones and count on the radio in Fido's collar to keep him in our yard.

Radio is one of the many applications of electronics. Others that come to mind are computers, entertainment systems, other telecommunications systems, control systems, security systems, manufacturing systems, automotive systems, game players and almost everything we touch or use in the twenty-first century. But radio was one of the first developments of electronic principles, because of its importance to public safety.

This book will explore radio—what it does and how it does it. *Basic Radio* will take you beyond the elementary electronic concepts presented in a book such as ARRL's *Understanding Basic Electronics* or a similar course of study in electronic principles to a more complete understanding of how radio works its magic.

David Sumner, K1ZZ
Executive Vice President
Newington, Connecticut
September 2005

Table of Contents

The national association for Amateur Radio

The seed for Amateur Radio was planted in the 1890s, when Guglielmo Marconi began his experiments in wireless telegraphy. Soon he was joined by dozens, then hundreds, of others who were enthusiastic about sending and receiving messages through the air—some with a commercial interest, but others solely out of a love for this new communications medium. The United States government began licensing Amateur Radio operators in 1912.

By 1914, there were thousands of Amateur Radio operators—hams—in the United States. Hiram Percy Maxim, a leading Hartford, Connecticut inventor and industrialist, saw the need for an organization to band together this fledgling group of radio experimenters. In May 1914 he founded the American Radio Relay League (ARRL) to meet that need.

Today ARRL, with approximately 170,000 members, is the largest organization of radio amateurs in the United States. The ARRL is a not-for-profit organization that:
• promotes interest in Amateur Radio communications and experimentation
• represents US radio amateurs in legislative matters, and
• maintains fraternalism and a high standard of conduct among Amateur Radio operators.

At ARRL headquarters in the Hartford suburb of Newington, the staff helps serve the needs of members. ARRL is also International Secretariat for the International Amateur Radio Union, which is made up of similar societies in 150 countries around the world.

ARRL publishes the monthly journal *QST*, as well as newsletters and many publications covering all aspects of Amateur Radio. Its headquarters station, W1AW, transmits bulletins of interest to radio amateurs and Morse code practice sessions. The ARRL also coordinates an extensive field organization, which includes volunteers who provide technical information and other support services for radio amateurs as well as communications for public-service activities. In addition, ARRL represents US amateurs with the Federal Communications Commission and other government agencies in the US and abroad.

Membership in ARRL means much more than receiving *QST* each month. In addition to the services already described, ARRL offers membership services on a personal level, such as the ARRL Volunteer Examiner Coordinator Program and a QSL bureau.

Full ARRL membership (available only to licensed radio amateurs) gives you a voice in how the affairs of the organization are governed. ARRL policy is set by a Board of Directors (one from each of 15 Divisions). Each year, one-third of the ARRL Board of Directors stands for election by the full members they represent. The day-to-day operation of ARRL HQ is managed by an Executive Vice President and his staff.

No matter what aspect of Amateur Radio attracts you, ARRL membership is relevant and important. There would be no Amateur Radio as we know it today were it not for the ARRL. We would be happy to welcome you as a member! (An Amateur Radio license is not required for Associate Membership.) For more information about ARRL and answers to any questions you may have about Amateur Radio, write or call:

ARRL—The national association for Amateur Radio
225 Main Street
Newington CT 06111-1494
Voice: 860-594-0200
  Fax: 860-594-0259
  E-mail: **hq@arrl.org**
  Internet: **www.arrl.org/**

Prospective new amateurs call (toll-free):
800-32-NEW HAM (800-326-3942)
You can also contact us via e-mail at **newham@arrl.org**
or check out *ARRLWeb* at **http://www.arrl.org/**

Chapter 1

Communicating Without Wires (Well, Almost...!)

Chuck Hutchinson, K8CH, works on the Yagi antennas on his tower.

Contents

Radio was on the logical evolutionary path behind a number of other systems designed to carry information over distances. Before there were electrical communications systems, there were a number of communications systems that were visually based. Anyone who has ever watched an American Western movie has likely seen Native Americans communicating through the use of smoke signals, only one of a number of visual-communications systems. Parallel developments going back perhaps even longer included the use of signal fires, as well as a sophisticated system of drum-based communications that could send messages across the African continent.

Semaphore

The insignia of the US Army Signal Corps, with which your author once served, consists of two crossed signal flags. These signal, or *semaphore flags*, could be used to signal over relatively large line-of-sight distances by the transmitting signalman holding the two flags at particular angular positions around the body, each position indicating a letter of the alphabet or a numeral. The space around the body was divided into eight positions with the first seven letters corresponding to left-hand flag down, with the right-hand flag progressing from part way up around the circle. The next seven with the left-hand flag part way up on the right, etc.

In an interesting bridge to later-day technology, rather than having separate codes for numerals, the code used the letters A through I to mean the numerals 1 through 9, if preceded by a special NUMERALS signal, in a manner similar to the much later Baudot teletype code. The positions of the flags for the letters A, B and C are shown in

Fig 1-1. While semaphore communication can be quite efficient, it suffers from some limitations. It can only be used in daylight, anyone within sight could read the mail and a signalman standing up on a high spot so his flags could be seen at a distance was an easy target. Perhaps that's why that kind of signalman is hard to find these days!

The mechanically driven semaphore, a device with mechanical arms activated by levers from an office below, and situated on a mountaintop tower was the basis for a communications network spanning much of Europe during the Napoleonic era, a long time before radio. These towers, positioned so each one could see the next, allowed a message to be relayed from station to station. While the data rate may have been slower than we're used to, the messages traveled at the speed of light! A variant of such a system, with electrically driven arms, was used as a railroad signaling system for many years and may still be in use in some areas.

Electrical Communications—the Telegraph

The *telegraph*, perhaps the first in a long line of electrically operated communications systems had its genesis in the invention of the electromagnet in 1825. A number of people operated an electromagnet from a remote location to perform rudimentary signaling, however, Samuel F. B. Morse is generally credited with developing the first complete system. This struggling artist and NYU Art Professor made a mid-life career change that changed not only his life, but the lives of everyone who followed as well.

His telegraph, first publicly demonstrated in 1838, was a major milestone and perhaps the first system that could be called a *tele-communications system*, although that term came into vogue many years later. The original Morse

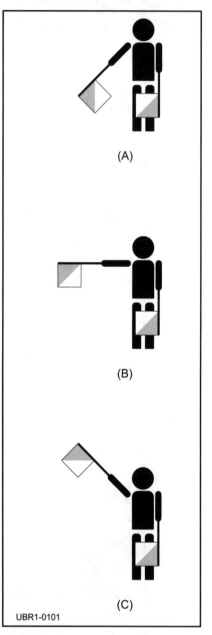

UBR1-0101

Fig 1-1—Hand semaphore positions for the letters A, B and C.

telegraph used a clever electromechanical arrangement to cause a pen to move across moving paper tape when the sending key was closed. This is the origin of the terms *mark* and *space*, the gap between marks, still in current use in data communications. Later versions used other marking mechanisms and eventually,

Fig 1-2—Early wire-line telegraph key and sounder. This was well used in US railroad telegraph service in the 1800s. (*Thanks to Dan Henderson, N1ND, for the loan of this family heirloom.*)

especially in the US, operators learned to more efficiently decode by ear, allowing a significant increase in speed and thus throughput.

In 1844, a government-sponsored demonstration system was put in place between Washington, DC, and Baltimore, Maryland. Following a successful demonstration to members of Congress the inventor and his partners extended the line to Phila-delphia and New York City and offered commercial service. In 1851, the telegraph began use as a railroad dispatch system, a telegraph key and sounder from the late 1800s is shown in **Fig 1-2**. Ten years later the Western Union Telegraph Company built a transcontinental telegraph system effectively making the two-year old Pony Express service obsolete overnight. Interestingly, the first transatlantic telegraph cable was installed some years earlier but burned out after only 30 days due to the fact that the users didn't fully understand the limitations of the then-current technology.

The Telephone— the Next Step

Alexander Graham Bell had been working on a *harmonic tele-graph*, a system that would allow multiple telegraph channels to be sent simultaneously over a single pair of wires, based on their being sent on different frequencies. Today we would call this a *frequency division multiplexer*, but at that time it led Bell to the thought that instead of carrying telegraph signals his apparatus might be used to carry the different sounds of human speech.

Bell succeeded in his experiments and in 1875 was able to file for a patent for his invention, surprisingly just hours ahead of another inventor. Bell's early telephones were offered as rental units, a practice that continued until the US antitrust settlement known as the *modified final judgment of 1984*. Renters of his apparatus would independently contract to have wires installed between the end points to provide a kind of intercom service. Within a few years after Bell invented his telephone, switchboards were devised to let a user communicate with other users, and in 1878 the first telephone exchange was established in New Haven, Connecticut. This made the telephone a viable facilita-tor of commerce and communica-tions. Long-distance service started quickly thereafter and for the next few decades, the history of the telephone was more a story of legal and regulatory battles between competing companies at each end of long-distance lines than of technol-ogy. The automatic (forerunner of the *dial*) exchange was invented in 1889 by Almon B. Strowger, a Kansas City undertaker, who was upset after finding that the local human telephone exchange operator was steering business to a competi-tor.

Next Step—Radio

While the telegraph was being rolled out across the US frontier and much of the rest of the world, a Scottish physicist and mathemati-cian, James Clerk Maxwell, was formulating the mathematical basis of electromagnetic theory. In 1861 he published his work predicting the waves that would travel from a wire carrying a changing current, laying the groundwork for what we call *radio*. Maxwell's equations allow the prediction of all the resulting fields involved in the generations of such waves.

It remained for another scientist, the German Heinrich Hertz in 1886, to actually demonstrate the existence of radio waves. His experiment was a validation of Maxwell's work, not an attempt to establish a new communi-cations medium.

Fig 1-3—Marconi shore radiotelegraph station at South Wellfleet, Cape Cod, Massachusetts in 1901. The towers were 200 feet tall and made of wood. They did not survive Atlantic storms for long. (*Photo courtesy of Marconi Corporation, PLC.*)

Development of radio into a commercial communications system, including a corporate structure that still bears his name, was left to Guglielmo Marconi, an Italian. Although Nikola Telsa is actually credited with the invention of radio, Marconi clearly developed radio into a significant business enterprise. He is said to have sent his first signals in 1895, communicated across the English Channel in 1899 and across the Atlantic three years later. His system used a variant of the Morse code then used by landline operators.

By that time, wire-line telegraph was well established for communications on land, and by undersea cable across bodies of water. The missing link was communication with ships at sea, which were at the mercy of the elements without any way to call for help. Marconi sought to fill that need through the provisioning of what today would be called a *turnkey* service. His company supplied the shore stations (see **Fig 1-3**), the shipboard stations (see **Fig 1-4),** and the operators for each. Shipboard radio operators wore the uniform of the Marconi Company. Their exploits at saving lives, often at risk of their own, made them heroes of the day.

Since the time of Marconi's radio operators, radio has grown in all dimensions and has become a major part of our lives in countless ways. We will explore these developments from both a technological and historical perspective as we progress through these chapters.

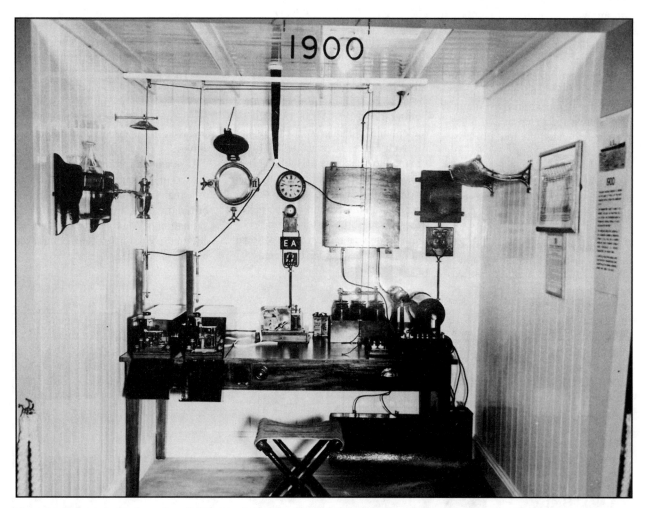

Fig 1-4—Marconi shipboard radiotelegraph station from 1900. (Photo courtesy of Marconi Corporation, PLC.)

Where do we start? It's a bit like the chicken and egg problem. How can we talk about receivers unless you start with transmitters, and vice versa? Every radio system contains a number of elements and we will cover each in turn. See **Fig 1-5**. While we could begin anywhere, here we will start in the middle with the actual "wireless" part of the system.

Radio seems a bit like magic. Fortunately we have all experienced this magic, so we start out as believers, making this a bit easier to explain. Your preparation in basic electronics should have introduced the concepts of *electric* and *magnetic fields*.[1] You likely learned[2] that a voltage applied to two metal plates generates an electric field between them, and that a current in a wire results in a magnetic field around it.[3] Radio happens because of the way changing electric and magnetic fields behave together.

You also learned that a change in the current in a wire results in a changing magnetic field, and that a wire in a changing field produces a voltage based on the strength of the field and the rate of change of the current. Further, a changing current in a wire will cause a changing magnetic field in the space around it, resulting in a changing electric field perpendicular to that magnetic field. The coupled set of electric and magnetic fields will move outward from the wire at the speed of light, forming what we call an *electromagnetic field*. The movement of this field is referred to as an *electromagnetic wave*.

While we can't see an electromagnetic wave resulting from a current in a wire, we can see such a wave when it is in the form of light. Yes, light is also an electromagnetic wave! This is handy, since most of us understand how light moves through space and how other objects interact with it. Electromagnetic radio waves share

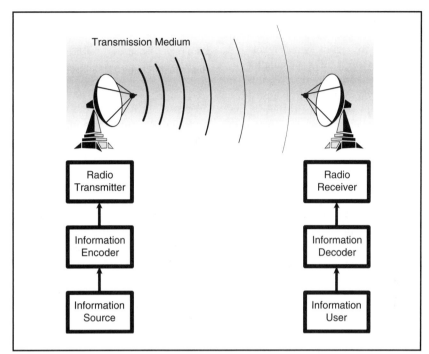

Fig 1-5—Elements of a radio communications system.

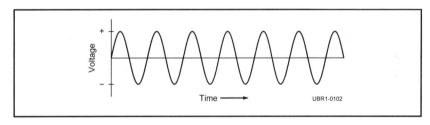

Fig 1-6—Sinusoidal signal waveform.

many of the same characteristics. The actions we observe with light—reflection, refraction, diffraction, absorption and filtering—all happen with radio waves as well. This makes radio somewhat easier to understand since we can relate it to something with which we are already very familiar.

How Do We Talk About Radio Waves?

We mentioned above that one of the requirements for an electromagnetic wave to be generated was a

time-varying current. A time-varying current can take a number of forms, but the most frequently encountered takes the form of a *sinusoidal*[4] waveform, as shown in **Fig 1-6**. Other waveforms can be used to generate an electromagnetic wave, but most practical systems are based on sinusoidal signals.

A sinusoidal signal can be described as having a *frequency*, the number of cycles it makes every second. If a sinusoidal signal is used to generate the time-varying current that results in a radio wave, then the wave will have the same frequency as the signal that generated it.

Frequency is a particularly important measure associated with radio waves since, as we will discuss, the signal's frequency often is a major factor in understanding how signals get from one place to another. Different frequency waves react differently with their surroundings.

A related measure of a wave is *wavelength.* Wavelength is just the distance the wave travels during one cycle at its frequency. This can easily be determined by dividing the speed that the radio wave travels (at the speed of light) by the frequency. So, for example, if we have a signal at 10 MHz, we just take the speed of light (300,000,000 meters per second) and divide it by 10,000,000 cycles per second to determine that the wavelength is 30 meters. Again, that's just how far the wave will travel during one cycle of its waveform. It's about one stone's throw!

Since the speed of light is constant, every signal could be described by either its frequency or its wavelength. In the beginning days of radio, no one could decide which to use so you will find early radios (before around 1930) with dial markings in both frequency and wavelength. Government regulations began to specify frequency rather than wavelength around then, so common usage locked in step and now frequency is almost always used to specify a radio signal. However, you will still find wavelength used to describe a band of frequencies, such as the 31-meter shortwave broadcast band or the 20-meter amateur band. This is convenient since when we talk about antennas, you will observe the relationship between wavelength and the physical size of antennas.

For purposes of discussion here, it is convenient to consider a grouping of radio frequencies. We will use the names in **Table 1-1** as we discuss the

Table 1-1

Name	Symbol	Frequency Range	Wavelength Range
Low Frequency	LF	30 – 300 kHz	10,000– 1000 meters
Medium Frequency	MF	300 – 3000 kHz	1000 – 100 meters
High Frequency	HF	3 – 30 MHz	100 – 10 meters
Very High Frequency	VHF	30 – 300 MHz	10 – 1 meters
Ultra High Frequency	UHF	300 – 3000 MHz	100 – 10 centimeters
Super High Frequency	SHF	3 – 30 GHz	10 – 1 centimeters

ways signals move from place to place.

Note that because 1000 kilohertz (kHz) equals 1 megahertz (MHz), we change notation as we move from MF to HF. Similarly, 3000 kHz is the same frequency as 3 MHz. We use MHz to avoid lots of zeros. In the same way, and for the same reason, we change from meters to centimeters as the wavelengths get shorter, as well as moving from megahertz to gigahertz (GHz). There are similar descriptions for higher and lower radio frequencies; however, the ranges shown in the table above include only those of interest to this book. Just to get a perspective on the various ranges:

• LF signals are used mainly for strategic military communication and some navigation systems (Loran C, for example, around 100 kHz).

• MF signals include the standard AM broadcast band (550 to 1700 kHz) and one amateur band, 160 meters. Before 1964 local ship-to-shore traffic was carried in this range; it has now been moved to the VHF range. The international shipboard radiotelegraph calling and distress frequency is in this range at 500 kHz. Until 1995, all US ships were required to monitor that frequency and have a licensed radio operator on-board who could communicate via radiotelegraph. The availability of reliable satellite systems is now considered sufficient to make the MF radiotelegraph

requirement obsolete. Radiotelegraph is still in common use in some parts of the world, however.

• HF is often referred to as *shortwave* and is the frequency range of many types of long-range communications, including international broadcasting, much of amateur radio, marine and aviation long-distance communications and others. Before the days of quality undersea cables for voice, HF carried most long-distance telephone conversations across the oceans.

• VHF carries the on-air VHF TV (channels 2-13), FM broadcasting, public-service vehicle communications, such as police, fire, taxi, local ship-to-shore, aircraft communication and navigation, plus some amateur segments.

• UHF is used for UHF television, cellular-phone service, military radar, *trunking* networks that carry multiple channels of public-service vehicle traffic and more amateur bands. The term *microwave* is used to refer to frequencies above 1000 MHz. This range includes city-to-city microwave telephone communication (mostly now replaced by fiber optical cables in the US).

• SHF is a continuation of the microwave region and is used for most satellite communications, some point-to-point terrestrial links and, no surprise, more amateur bands!

So How Do the Signals Get There?

One of the facets of radio that make it an interesting topic is the manner by which signals get from one place to another. Signals can always be described by the *line-of-sight* path that works just like light. It starts here and goes there, and you can see one point from the other. How far you can shine a light depends pretty much on how bright the light is and how sensitive the eye is. So it is with radio signals.

A path is pretty easy to determine with just a bit of analysis. You must know the height of each endpoint and how much the earth curves in-between to determine the maximum line-of-sight distance. If there are obstructions in the path, it is no longer line-of-sight and we have to look for other means to get the signal around the obstruction. As with light, radio signals can reflect off objects (especially metal, but also water, for example) and refract (bend) while traveling through some objects, just as light does going through water.

The intensity of the light—or of a radio signal—as it moves from the source to the observer appears to diminish as the distance gets greater. This is true even if the signal is not being absorbed by anything in between, but is due to the amount of energy available to your eye—or to the receiving antenna—due to the energy spreading out over a wider area as it travels from the source.

Just as light can be reflected from a mirror, radio waves can be reflected, but by an *electric mirror*. The actions of such a electric reflector can be very much like those that reflect light. For example, the reflector behind the bulb in a flashlight or car headlight has a shape similar to the reflector behind an aircraft's radar antenna. The difference is that the radar reflector is made from a conducting material such as metal, while the light reflector is usually polished metal, or it could be a non-conductive material with a shiny surface.

If you look at various radio reflectors, you will notice that they aren't always solid—sometimes they are made of screen-like material. If these were used as reflectors for light, some of the light would leak through. Why doesn't this happen to radio waves? The reason is that light is at a very much higher frequency (a shorter wavelength) than radio frequencies. The designer of a radar reflector selects the size of the holes in the reflector so that they would be small in terms of the radio signal's wavelength and thus the radio signal doesn't see the holes at all. We can consider a reflected signal as being a *bent* line-of-sight signal.

Something that happens with lower-frequency radio waves that doesn't happen with light is called *ground-wave propagation*. Here the radio signal tends to be absorbed as it passes over the ground, resulting in the signal being *tilted*. Think of it being dragged down as it travels along. The dragged-down signal hugs the ground and continues along to provide coverage beyond the line-of-sight distance. The signals get weaker quickly, however. This is the way high-powered AM broadcast transmitters reach radios 50 to 100 miles away during the day, far beyond the line of sight.

So What Happens To Signals That Go Up?

Signals that leave the transmitting antenna in an upward direction are called *sky wave signals*. They can travel in a line-of-sight fashion, such as in ground-to-air communications, but the most interesting ones to hams are the HF signals that interact with something called the *ionosphere*. The ionosphere is a region surrounding the earth at a height of roughly 30 to 260 miles (see **Fig 1-7**). Because this

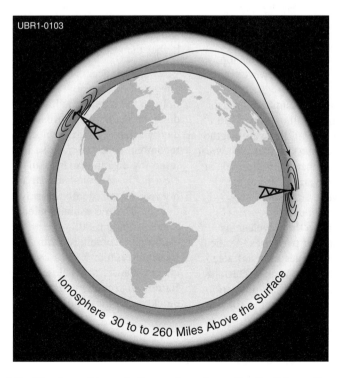

UBR1-0103

Ionosphere 30 to to 260 Miles Above the Surface

Fig 1-7—The ionosphere is the region above the Earth's surface.

is farther away from the pull of gravity than our atmosphere, the molecules at ionospheric heights are less dense than they are where we can breathe. The density of air molecules is reduced from the bottom of the ionosphere towards the upper reaches, until there are almost no molecules left, just as in deep space.

Since the effect on radio waves varies with the height of the iono-sphere, it is convenient to talk about the ionosphere in terms of *regions,* also known as *layers,* as shown in **Fig 1-8**. In all portions of the ionosphere, energy from the sun during daylight periods interacts with the molecules. Ultraviolet radiation has the biggest effect but X-rays and other forms of radiation interact with the air molecules as well. The effect of radiation is to raise the energy in the molecules, causing them to release electrons. When an electron is released from an electrically neutral molecule (that is, one that has the same number of positive and negative particles) the result is a free electron (negative charge) and a positively charged *ion.*

A radio signal encountering an ionized area will react in one of the following ways, depending on its wavelength and the density of ions:

1. The radio wave can be ab-sorbed. This will occur if the ion density is high and the wavelength is long.

2. It can be transmitted through the ionosphere without much loss. This will happen if the wavelength is short compared to the space between ions. It also depends on the angle at which the wavefront strikes the ionized layer.

3. It can be bent, or *refracted.* This happens to the wavelengths between the first two cases. As the wavefront continues to be refracted, it tends to bend down and return to earth. Even though it is a refraction rather than a reflection, the effect is

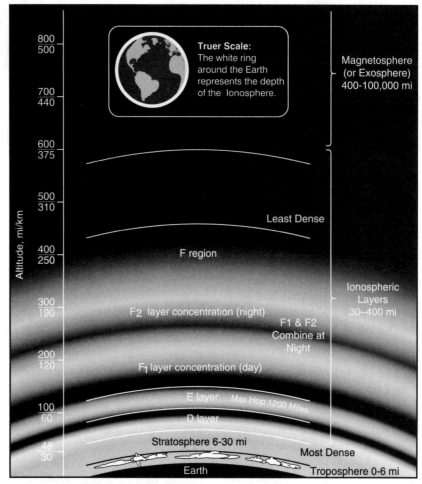

Fig 1-8—The ionosphere considered as layers

similar to a reflection from a surface higher than the actual refracting area.

Ionization occurs because of the energy of the sun, and it happens only on the side of the Earth in daylight. As the Earth rotates from daylight into night, free electrons recombine with the positive ions, returning air molecules to an electrically neutral state. All areas do not recombine equally quickly though. The recombination occurs in a rather random fashion with electrons recombining with ions they happen to run into. As a conse-quence, denser areas closer to the Earth's surface recombine more quickly, and those further away take

longer, sometimes staying ionized throughout the night.

The combination of these effects and the variability of solar activity have made HF radio propagation difficult to predict in the past. In the computer age, sophisticated com-puter modeling programs have made it possible to predict the frequencies to use to maximize the likelihood of successful transmission in any of the above modes. Of course, actually trying out each path at a number of frequencies is an even better method and some automated HF radio systems do exactly that, simulating the knowledgeable radio operator of days gone by, but in seconds instead of hours!

Any piece of wire or metal tubing could work as an antenna if we could cause a time-varying current to flow in it. The trick is to get the current to flow properly. While there are many exceptions and special cases, this is easiest to do with a structure comparable in size to half a wavelength at the frequency we want to launch.

It is important to not confuse the *antenna* with the *support structure*. For example, the antenna for a MF AM broadcast station is likely a vertical tower from 150 to 500 feet tall. In this case the tower *is* the antenna. A VHF-FM broadcast station will have an antenna that is only 10 to 20 feet in height, but to obtain a long line-of-sight path to its listeners, it may be located on top of a tower or building that is 1000 feet high. Unless you look carefully it may be hard to tell the difference between antenna and support structure!

The Dipole Antenna

The simplest and one of the most frequently encountered types of antenna is called a *half-wave dipole* and is shown in **Fig 1-9**. It is merely a length of wire or metal rod an electrical half-wavelength long, generally split in the center and connected to a transmitter sending signals on a frequency corresponding to the half wavelength. In many cases it is desirable to have the transmitter and antenna located in different places, connected by a *transmission line*, which we will discuss further in Chapter 16.

It is also possible to use half of a dipole as an antenna. This is often employed when the antenna is vertical (as in the AM broadcast antenna tower mentioned above). Essentially, the missing half of the dipole is *mirrored* by a connection to a ground system beneath the antenna. Such an antenna is often called a *quarter-wave monopole* antenna and its effectiveness is directly related to the quality of the ground system.

While a half-wave dipole or quarter-wave monopole is easy to sketch and discuss, at lower frequencies neither may be practical. Consider an AM broadcast receiving antenna. At a frequency of 1 MHz, near the middle of the AM band, the wavelength is 300 meters. A half-wave antenna would be 150 meters, or almost 500 feet long. A quarter-wave monopole would be 250 feet long. These lengths are practical in some applications, but 250 feet would be rather long on a car fender!

More Complex Antenna Systems

It is quite possible to construct antenna systems using more than a simple dipole. Earlier we discussed how a reflector behind a light bulb makes the light appear stronger in some directions and weaker in others. Antennas work the same way. The energy leaving an antenna can be focused through an appropriate *lens* or reflector. A reflector like that of a flashlight can be effectively used at UHF and higher frequencies, where it can be a manageable size. At lower frequencies it is more common to use less-complex reflectors or multiple-dipole elements to focus the signals. We will discuss the ionosphere and antennas more when we get to Chapter 16.

Notes

[1]Wolfgang, L, *Understanding Basic Electronics*, ARRL, Chapter 9. Available from the ARRL Bookstore for $20 plus shipping. Order number 3983. Telephone toll-free in the US 888-277-5289, or 860-594-0355; **www.arrl.org/shop/ pubsales@arrl.org**.
[2]Ibid, Chapter 17.
[3]Ibid, Chapter 19.
[4]Ibid, Chapter 16.

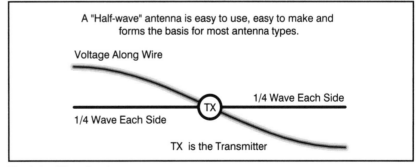

Fig 1-9—The dipole antenna.

A "Half-wave" antenna is easy to use, easy to make and forms the basis for most antenna types.

Voltage Along Wire

1/4 Wave Each Side

1/4 Wave Each Side

TX

TX is the Transmitter

Review Questions

1. Radio waves act a lot like light. In fact, before radio was around, light was used as a mechanism for medium-to-long distance communications. Can you think of three types of early light-based communication systems?

2. Compute the approximate length of a "half-wave dipole" antenna designed for the AM broadcast band, say 1 MHz; the 31-meter international broadcast band at 9.5 MHz; TV channel 2 at 58 MHz; and a satellite earth station downlink at 3000 MHz. What do you notice about the size of these antennas?

3. Consider a world without an ionosphere. How would radio communication be different than in our world?

Chapter 2

The Simplest Receiver
—the Crystal Set

An early crystal radio using a "Cat's Whisker" galena detectorl

Contents

As was the case in Chapter 1, we have another chicken and egg problem—how do you describe a receiver if you don't know what kind of signal you will be receiving and the type of transmitter that sent it? As before, I will introduce just enough about transmitters and signals to be able to start a discussion on receivers.

Signals—What Do We Mean By Signals?

In Chapter 1 we discussed radio signals and how they move through space. We talked about alternating current (ac) signals as sinusoidal voltages and currents resulting in electromagnetic waves moving outward from antennas. A single-frequency ac signal will certainly do exactly that, but it is not a very interesting signal unless it carries *information*. Sometimes just being there is all the information we need, such as would be true of a primitive rescue beacon. But most of the time we want the signal to carry voices, music, pictures or data.

How Can We Apply Information To a Signal?

Each sinusoidal signal can be described by a number of parameters:

- *Frequency*—This is the number of complete cycles the signal makes per second.
- *Amplitude*—Although the amplitude, or strength, of a sinusoid is constantly changing with time, we can characterize the amplitude by the maximum value that it reaches.
- *Phase*—The phase of a sinusoid is a measure of when it starts compared to another sinusoid at the same frequency.

Any of the above parameters can be changed to apply information to the signal in a process called

modulation. A crystal set can receive *amplitude-modulated signals*. The process of transmitting amplitude modulation that a crystal set can receive simply requires that we multiply a signal by a waveform conveying the information. An example should make this easy to grasp. For our signal, we will again consider the sinusoid that was shown in Chapter 1 in Fig 1-6. Let's say we want to transmit the letter "a" in the ASCII computer language. The lower case "a" is represented in computer language (ones and zeros) as 1100001. The next step in the process is to convert that code into a waveform. We will represent the "ones" by pulses of 1.0 V and the "zeros" by pulses of 0.0 V. The data (or information) waveform looks like **Fig 2-1**.

We now multiply the two waveforms. During the time we have a 1-V pulse we see a transmitted *carrier wave*. During the time we have a zero for data, we have no output. We could send this code across the street by turning a light switch on and off! If we want to use radio waves instead, the carrier can be turned on and off, and we have the simplest form of amplitude modulation, as shown in **Fig 2-2**.

How Do We Get the Information Out at the Far End?

If the transmitted waveform reaches us, we can *detect* it with an antenna connected to a simple crystal receiver. The simplest receiver circuit is based on a half-wave rectifier[1] and it will form the first basic building

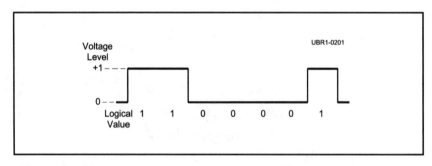

Fig 2-1—Voltage pulses corresponding to the ASCII representation for the lower-case letter "a".

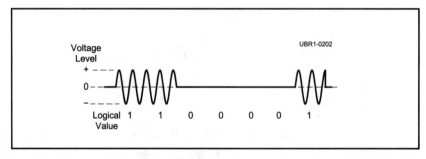

Fig 2-2—Sinusoidal waveform modulated by the data waveform of Fig 2-1.

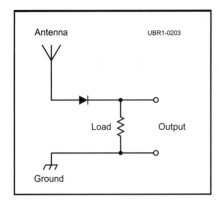

Fig 2-3—The simplest radio receiver. For this type of receiver to work, there must be a dc path between the antenna and ground.

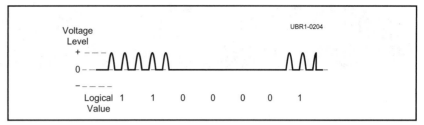

Fig 2-4—Received pulsed-data waveform from a simple receiver.

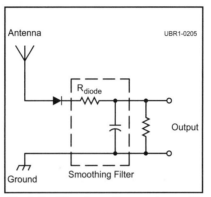

Fig 2-5—Simple radio receiver with added smoothing filter.

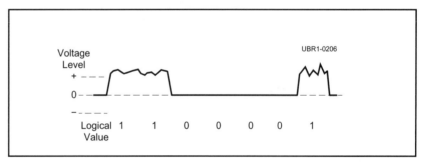

Fig 2-6—Received pulsed data waveform from receiver with smoothing filter.

block of a simple receiver. If we apply a modulated signal to a half-wave rectifier, with the input a connection to an antenna and ground (**Fig 2-3**), the voltage at the output is a series of positive carrier pulses during the *ones* and zero voltage during the *zeros*, as shown in **Fig 2-4**.

The signal doesn't look quite like the data we are trying to recover, but by adding a carefully selected capacitor value—one that will charge and discharge between the cycles of the carrier but not stay charged during the zeros—we can recover a reasonable copy of our original data. This capacitor is referred to as a *smoothing filter*. A receiver with a filtering capacitor is shown in **Fig 2-5**, with the resulting data output in **Fig 2-6**. While this isn't quite a perfect copy of the original, a computer could easily recognize the data.

Yes, we could actually build our own crystal radio receiver! Retailers used to sell electronic parts in most towns and cities, but today it is a bit more difficult to find electronic parts needed for projects like this. There are many computer parts available, but not as many radio parts available as there used to be.

So What Do We Need?

Luckily, as noted in Fig 2-5, there aren't very many parts required for a simple receiver! It may, however, be hard to find the high-impedance headphones that were popular when crystal sets were "high tech" back in the 1920s. The values for the other parts are not too critical, but should be fairly close to the values shown.

For the crystal we can avoid the difficulty of finding old-fashioned galena crystals and buy a very inexpensive modern semiconductor diode. It will work better and needs no adjustments. Almost any diode will work, but your radio will be more sensitive with a germanium diode, such as a 1N34, than with a more common silicon diode, such as a 1N914.

Construction methods are not too critical either. Perhaps the best way to build this is by placing the parts on a perforated circuit board and soldering the connections. Use the minimum amount of heat on the diode to avoid damage. Use just enough heat to solder to its leads.

The most effective antenna we can probably manage for this simple radio is a piece of wire as long and as high as practical. If we really wanted to do the best job we could, we would want to have a vertical wire $1/4$ wavelength long. This makes for half a *dipole* and requires a good ground for the other side of the connection. Look at your answers to Review Question 3 in Chapter 1 and divide by two for a $1/4$-wave antenna. But this is probably still longer than you can deal with. So your antenna

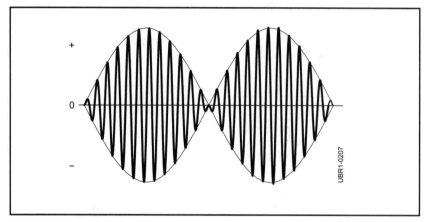

Fig 2-7—A single tone amplitude modulating a carrier. In this example, there would be 600 cycles of the carrier in each cycle of the modulating waveform. The recovered audio waveform is a 600-Hz sine wave.

will be shorter than optimum, and will therefore pick up less signal. As a young lad, I had good results with about 30 feet of wire run to a tree outside my second floor bedroom window, perhaps 10 to 15 feet above ground. I was able to hear strong broadcast stations from about 25 miles away.

For a ground connection, the classic connection point in a house is a cold-water pipe. This works because pipes are usually copper and they end up underground outside the house. Make sure there's no PVC (plastic) pipe in the path, especially if your house was built fairly recently. Plastic pipes do not conduct electricity! Another ground possibility is to connect to the screw on the outside of a power outlet plate. If proper grounding has been employed in your house's wiring, this should connect all the way back to the ac mains service-entrance ground.

How well you can receive stations with this simple receiver will depend on how close you are to an AM transmitter, the power of that transmitter and the length and the effectiveness of your antenna and ground connections.

How Do I Hear the Top-40 with my Simple Radio?

Yes, a simple receiver can do a nice job picking up AM broadcast signals. A voice (or music) AM signal works pretty much the same way as the data signal illustrated in Fig 2-2. Instead of multiplying a transmitter carrier signal using ones and zeros, we can modulate with a more complex waveform. Suppose we have a carrier signal of 600 kHz and wish to multiply it by the signal from a musical instrument holding a steady note of a single frequency, say 600 Hz. The resulting waveform would look like **Fig 2-7**.

A simple crystal receiver like Fig 2-5 will recover a slightly distorted copy of the 600-Hz tone, which we could listen to by replacing the load resistor with sensitive, high-impedance earphones. While the original audio signal is centered around 0 V, the recovered audio signal from our simple crystal receiver is centered around a dc voltage equal to half the peak voltage of the received carrier. This causes a slight bias to the position of the earphone diaphragm, and can be removed by inserting a series

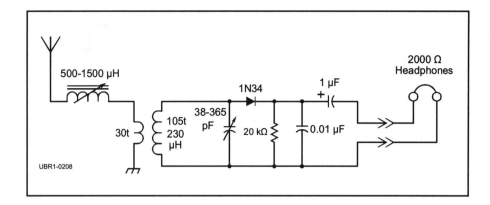

Fig 2-8—Schematic diagram of a crystal receiver with dc blocking capacitor. Note that a series coil is used to resonate the capacitive reactance of the antenna. Further, the antenna is transformer-coupled to the input tuned circuit to avoid undue loading, which would decrease receiver selectivity.

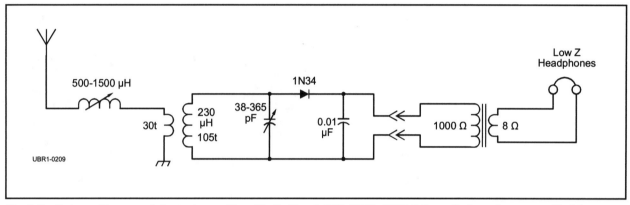

Fig 2-9—Schematic a crystal receiver with a RadioShack 273-1380, 1000:8-Ω audio transformer so that low-impedance headphones can be used.

capacitor to block the dc component of the signal, as shown in **Fig 2-8**.

What's Wrong With This Simple Receiver?

A receiver like this actually does work and I encourage readers to try it. If you do, you may find out that it has some significant limitations. The following are the most noticeable:

1. It will receive not only the signal we want, but also all others at the same time. This is not a problem if the signal we want is far stronger than all others, but in the real world we will need *selectivity* to receive just our desired signal. Receiving five top-40 stations at the same time is not necessarily an improvement over receiving only one!

2. This receiver puts out a signal equal to the received carrier level. It is actually *powered* by the energy received from the transmitter, but if we're not close enough to the transmitter or don't have a big-enough antenna, we will struggle to hear anything. There is also no way to easily control the received level. We will need adjustable *amplification* or *gain* to overcome this limitation.

3. Another limitation mentioned previously is that most modern headphones are low impedance, typically 600 Ω or lower, sometimes as low as 8 Ω. The earphones of the crystal-set era back in the 1920s were 2000 Ω or higher, designed to give higher sensitivity for such

applications. Lower-impedance phones either will provide less output or will require a stronger signal. A high-impedance to low-impedance output transformer can help avoid this problem, as shown in **Fig 2-9** in which a 1000:8-Ω audio transformer (RadioShack 273-1380) has been added so that low-impedance headphones can be used.

The next sections will describe more details of how we can overcome these limitations. If you built a simple receiver on a breadboard and left some room you will be able to easily make these improvements.

From a Historical Perspective

The steps towards improving the simple crystal set that we're going to discuss in the following sections are pretty much the way radios evolved in the early part of the 20th century. The earliest receivers were very much like our simple crystal set. Instead of a packaged semiconductor diode, however, early radio pioneers used a piece of galena crystal with a *cat's whisker* wire probe to make their crystal detectors. This was before vacuum tubes were developed, so amplification was not an option. The pioneers first worked on the selectivity problem, refining it to quite an art with various types of exotic tuning mechanisms intended to bring in only the signals they wanted.

Fig 2-10—Close-up of a coherer tube in a Marconi detector. *(Photo courtesy of the Marconi Corporation, plc.)*

There was Other Technology Too

The crystal set is the kind of radio many would associate with the earliest radio equipment, but it may have survived into current literature largely because it is easy to duplicate with available materials and has the virtue of receiving voice signals as well as the radiotelegraph signals for which early receivers were designed. There were earlier and competing technologies as well.

The Coherer Detector

The early Marconi stations made use of a device called a *coherer*, named after the action of metal filings under the influence of a voltage across them. If a voltage is applied to a collection of loose metal filings, they tend to line up between the charged electrodes in response to the electric field between them. A phtograph of a coherer detector tube is shown in **Fig 2-10,** with a view of the complete Marconi coherer in **Fig 2-11**. Loose filings show a high

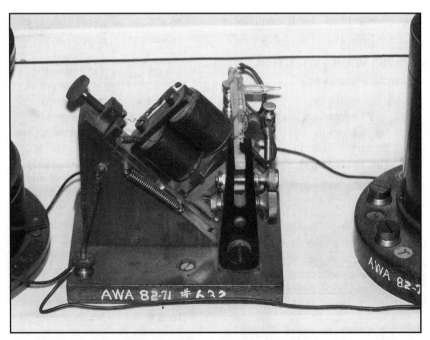

Fig 2-11—Complete Marconi coherer detector. *(Photo courtesy of the Antique Wireless Association.)*

resistance, while lined-up ones exhibit a low resistance, allowing current to flow in a secondary circuit. The electrodes in the partial vacuum in the coherer tube were adjustable in-and-out to give maximum sensitivity. This occurred at the point just before the filings were pushed into direct contact.

The coherer receiver worked with a second circuit operating a relay. The relay could be used to provide a local signal to a pen recorder of the type used in land telegraph, allowing an automatic record to be made, or to a sounder, as if the radio system became a direct replacement for the connecting wires of a land-telegraph system.

Unfortunately, the coherer did not *decohere* by itself after the radio pulse had finished. That is to say that once the filings lined up they tended to stay where they were until the coherer tube was tapped or vibrated to shake them back to their rest position. Fortunately, this was another function of the second circuit, in that the relay could be provided with additional contacts that caused a bell-like clapper to strike the tube to shake the filings free. The other limitation of the coherer was that it took quite a strong signal to move the filings in the first place. Many early receivers were equipped with multiple detectors so that the optimum one for the conditions could be selected. A photo of a Marconi receiver with coherer detector is shown in **Fig 2-12**. The coherer was, in effect, an on-off switch, not a detector that could recover AM voice signals. It thus never became popular with the residential market. Another disadvantage was that occasionally the decoherer hammer would break the glass tube rather than just shaking the filings loose. Of course, this only happened when a message was being received.

The Magnetic Detector

An unusual detector attributed to Marconi was the *magnetic detector*. This device consisted of an endless band of stranded, silk covered iron

wire that was passed by permanent magnets and an input and output coil. The signal was applied to the input coil and the headphones were connected to the output coil. The wire was pulled past the magnets and

the coils with a clockwork mechanism.

Marconi found that the magnetic effect of the permanent magnets upon the wire could be distorted by the application of an ac signal to the

Fig 2-12—A complete Marconi coherer receiver. *(Photo courtesy of the Marconi Corporation, plc.)*

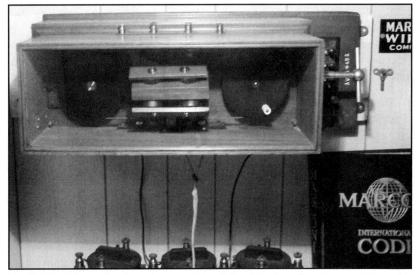

Fig 2-13—Close-up of a Marconi magnetic detector. *(Photo courtesy of the Antique Wireless Association.)*

wire as it moved through the assembly. This would result in a signal being heard in the headsets. **Fig 2-13** provides a closeup of a magnetic detector.

The magnetic detector was more sensitive and reliable than the coherer detector and remained the standard detector on Marconi equipment until the crystal detector was developed. An early Marconi receiver with magnetic detector connected to tuning circuit is shown in **Fig 2-14**.

The Crystal Set

The crystal set made the earlier detectors obsolete. It had more sensitivity than its predecessors, was less complicated to manufacture and could detect later voice transmissions as well as the telegraph signals of the day.

The crystal set had its drawbacks as well as advantages. Although a crystal set was perhaps the most sensitive of the early radio detectors, it only operated at all if the detector's cat whisker was positioned on a sensitive spot on the crystal material, a delicate and fragile proposition. The exact spot was hard to locate without a signal present, although an auxiliary buzzer could be used as a substitute signal for adjustment and was part of the Marconi equipment. Heavy seas or shellfire could result in the whisker being dislodged, with consequent loss of reception. In addition, crystal detectors were vulnerable to the kind of pulses that lightning or static discharge could generate, a real problem on shipboard.

Enter the Vacuum Tube— Modern Electronics is Born!

In the days before the US entry into WW I, the Deforest Company marketed the Audion,[1] a vacuum tube with versions designed to provide both *detector* and *amplifier* functions. A diode detector had the advantage of sensitivity almost comparable to a well-adjusted crystal detector, but it required no adjustments and neither did it show other quirky traits. On the negative side, however, the vacuum-tube detector was expensive, required

Fig 2-14—Complete Marconi receivering setup with magnetic detector and multiple tuner, circa 1910. *(Photo courtesy of the Antique Wireless Association.)*

battery power, was fragile and made for a less portable receiver than a crystal set.

The vacuum tube was more successful early on as an audio amplifier following a crystal detector. This gave enough audio signal power to allow a family to listen around a loudspeaker, rather than one person at a time listening to headphones. We'll talk about amplifiers in the next chapter, first on the audio side of the detector and then on the radio frequency side. As noted in the following sections, there was a lot of room for improvement before we could say we moved on to "modern electronics."

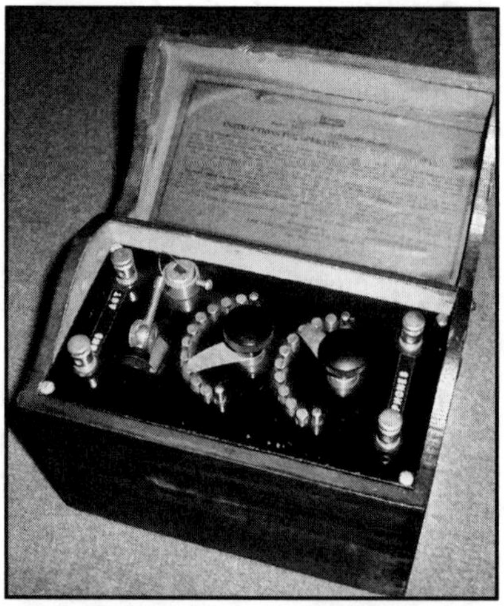

A Step Towards Better Selectivity

A *parallel-resonant circuit*[2] exhibits its maximum impedance (and thus highest voltage) at its resonant frequency. If we put a signal from an antenna across such a circuit and *tune* the circuit to be resonant at the frequency of the station we want to hear, that station will be emphasized more at the detector compared to stations at other frequencies. An early crystal set with a tuning circuit is shown below in **Fig 2-15**.

Fig 2-15—Early crystal set with built-in tuning circuitry.

Note that to be able to select different stations, we need to be able to change the value of either the inductor or capacitor, or both. In the early days it was usually the inductor (then called a *coil*) that was variable. Later, the capacitor (then called a *condenser*) was more often the variable element.

Let's see if we can design a resonant circuit that will cover the standard AM broadcast band using a variable capacitor. The resonant frequency of a parallel-tuned circuit[3] is:

$$f_r = \frac{1}{2\pi\sqrt{LC}} \qquad \text{(Eq 2-1)}$$

Note that there are no "magic values" used here. Any combination of inductor and capacitor values can be selected to meet the above resonant condition at the desired frequency. If we want to cover the standard AM broadcast band, from 550 to 1700 kHz, we will be covering a frequency range of 1700/550 or about 3.1:1. Note from Eq 2-1 that the frequency varies with the square root of the capacitance, so we will need a variable capacitor that covers a range of values of about 9.6:1, the square of 3.1:1. A standard maximum value used for broadcast coverage is a 365-pF variable capacitor, with a minimum value (including wiring capacitance) of capacitance no more than 38 pF, a practical value.

Knowing the maximum value of capacitance, and noting from Eq 2-1 that the larger the capacitance the lower the frequency, we can determine the value of inductance required to make the circuit resonant. We solve Eq 2-1 for L in terms of C and f with the following result:

$$L = \frac{1}{C(2\pi f_r)^2}$$
$$= \frac{1}{365\,\mathrm{pF}\,(2 \times 3.14 \times 550\,\mathrm{kHz})^2} = 229.6\,\mu\mathrm{H}$$

$$\text{(Eq 2-2)}$$

As a check we can plug the values of L and C back into the resonant frequency formula and we should get a frequency of 550 kHz. If we use 229.6 μH and 38 pF (the minimum value for the capacitor) in the resonant-frequency formula, we confirm the upper end of the frequency range at 1700 kHz.

In most radios, the inductance is made with a moveable iron slug to *trim* the value so that it is resonant at the right place on the dial, compensating for any stray reactive components in the circuit wiring and for the effect of the antenna.

Hooking up the Antenna

Having a selective resonant circuit should help us be able to separate stations. However, we must use care to keep from diminishing selectivity when we connect the antenna. There are two concerns. One is a change in resonant frequency and the other is a reduction in selectivity due to loading or reducing the *Q* (quality factor) of the circuit.

All wiring has both stray capacitance and inductance, in addition to any resistance in the wire itself. These effects can be determined by calculation based on the physical properties of the wires and their surroundings. For the wiring in a small radio, particularly at broadcast-band frequencies, the ability to trim the inductance in the resonant circuit means we needn't be unduly concerned about the actual values of strays, since they are quite small. The antenna connection is quite another matter. At broadcast-band frequencies, the usual outside wire antenna needed to pick up signals for our crystal set is not itself resonant, as we discussed earlier. Our "random-wire" antenna looks like a moderate-sized capacitor shunted by a large resistor.

If we were to hook such an antenna directly to the resonant circuit we would find that the resonant frequency shifts down due to the additional parallel capacitance.

The effect of the shunt resistance is to lower the Q to the point that it might no longer have sufficient selectivity to separate stations.

Instead of directly connecting to the resonant circuit we can overcome these issues by using transformer coupling to match the antenna impedance to the radio's resonant circuit and by using a series coil to cancel the antenna's capacitive reactance. These two techniques were shown in Figs 2-8 and 2-9 earlier.

Note that we now have two circuits to adjust and this is just what people did in the early days. **Fig 2-16** is a photo of an early two-coil tuner for a crystal set designed to optimize both tuning and antenna coupling.

On the other side of the tuned circuit, we have another potential problem. The headphones act like a resistive load on the tuned circuit and also lower the Q. This is another reason for using high-impedance headphones to listen to a crystal set.

Notes

[1]Wolfgang, L, *Understanding Basic Electronics*, ARRL, Chap 26, Figure 7. Available from the ARRL Bookstore for $20 plus shipping. Order number 3983. Telephone toll-free in the US 888-277-5289, elsewhere 860-594-0355; **www.arrl.org/shop/ pubsales@arrl.org**.
[2]Ibid, Chap 24.
[3]Ibid, Chap 24, p 24-7.

Fig 2-16—Photo of an early two-coil tuner for enhanced selectivity.

Review Questions

1. Assume that the variable capacitor for a crystal set is linear—that is, the capacitance increases the same amount with each equal amount of rotation. What would the effect be on the frequency dial? For example, consider a minimum of 38 pF at 0° rotation and a maximum of 365 pF at 180°. Then 119.75 pF would be at 45°, 201.5 pF at 90°, and 283.25 pF at 135°. Calculate the corresponding frequencies at each of the five positions. What do you notice about the dial scale? What could be done to improve the "tunability"?

2. If a radio's tuned circuit bandwidth happens to equal 5% of the resonant frequency, what would the bandwidth be at each end of the radio dial? What would be the impact of that be on the ability to separate stations?

3. Compute the length of a half-wave dipole for each end of the broadcast band.

Chapter 3

Enhancing the Simplest Receiver

An early Heathkit 3-tube radio receiver kit.

Contents

With the addition of a selective resonant circuit at the front of our receiver, we can start to look at the back end and address some of the other concerns raised in Chapter 2. We will not consider the selectivity question closed, since we will improve upon it in other ways later in this chapter, but we will move forward in a manner that parallels actual early receiver development.

An Audio Amplifier—Just What We Need

Radio began to have the capability to fill a room with sound around the time of WW I with the advent of the triode vacuum tube.[1] The development of the triode[2] is a worthwhile topic for those with an interest in the history of technology. There are still applications that are best served by vacuum tubes, especially high-power radio-transmitter stages using special tubes for UHF and microwave radar. Nowadays, however, solid-state devices more readily handle most receiver functions, so we will focus on their use in this book.

In our original crystal set described in Chapter 2, the only ways to increase the volume of a received signal were to extend the antenna or move closer to the transmitter. Both of these solutions have obvious practical limitations! So the introduction of amplifying devices made radio a more practical medium for entertainment and information distribution.

A triode vacuum-tube amplifier uses a small signal applied to the control grid to change the current in the circuit between the plate and the cathode. See

Fig 3-1. A small change in grid voltage from the AC Input Signal (such as a crystal detector) results in a larger change in plate current and the resulting signal coupled from the plate circuit is an amplified version of the signal on the grid. The result is a louder signal from the crystal detector. Multiple amplifier stages can build up the signal to any desired level.

With enough audio amplification, we can take the weak signal from a crystal radio and increase its level to fill a room. One way to easily verify this is to hook the output of your crystal set into the auxiliary input of a stereo system and turn up the volume. In the early days of "Hi-Fi," crystal-set AM tuners were a popular accessory, since they did not introduce any extraneous hum or noise into the system from the circuits associated with the vacuum tubes of the day. They were most popular with listeners who lived in major metropolitan areas near big transmitters.

How Do We Make It Happen?

While we can still find tubes, for our receiver we will use a transistor amplifier.

After all, transistors are much easier to find and they avoid potentially lethal voltages. We will add a simple single-transistor audio amplifier[3] between the output of our crystal receiver and our high-impedance headphones. In the circuit shown in **Fig 3-2** I have added such an amplifier to the crystal radio and included a volume control between the stages. This might be a bit optimistic, but I put it there so that the output level can be reduced if it is too loud.

This amplifier provides a voltage gain into a high-impedance load, such as the crystal-set headphones discussed in the last chapter. You could modify the amplifier to drive a pair of the more common low-impedance headphones by changing to a common-collector configuration, as shown in **Fig 3-3**.

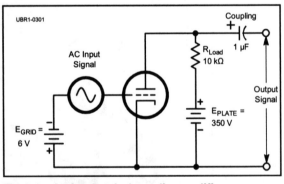

Fig 3-1—A simple triode audio amplifier.

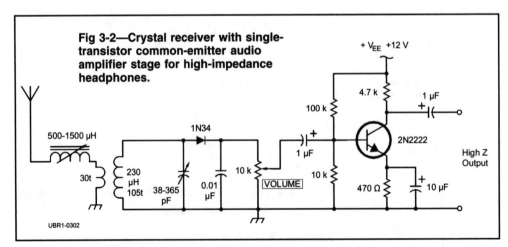

Fig 3-2—Crystal receiver with single-transistor common-emitter audio amplifier stage for high-impedance headphones.

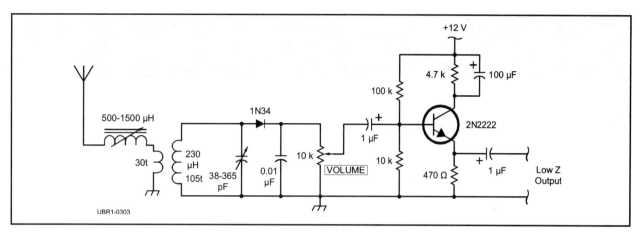

Fig 3-3—Crystal receiver with single-transistor common-collector audio power amplifier stage for low-impedance headphones.

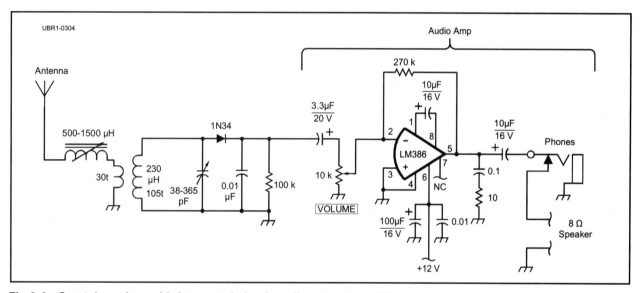

Fig 3-4—Crystal receiver with integrated-circuit audio output amplifier.

The common-collector amplifier is often referred to as a *power amplifier* rather than the usual *voltage amplifier* shown in Fig 3-2. A voltage amplifier provides an increase in the voltage of a signal, operating into roughly the same load impedance as seen by the source. A power amplifier typically provides an output at about the same voltage as the input, but at a higher current so that the power is higher. This is perfect to drive low-impedance headphones or even a loudspeaker.

Moving Into the Integrated-Circuit Age

With very little additional effort, we can make use of an *integrated circuit* amplifier stage. See **Fig 3-4**. An integrated circuit is a device that includes many transistors and other components designed to provide a completely functioning circuit. They are available both in the form of *analog* circuits and *digital* ones. Digital integrated circuits are the reason that computers fit on a desk instead of taking up rooms as they used to do back three or four decades ago.

The LM-386 analog integrated circuit I've specified includes stages internally that provide both gain and sufficient power output to drive a loudspeaker. If you have already built the single-transistor amplifier, it can be used as a *preamplifier* stage between the detector and the integrated circuit.

This circuit makes the crystal receiver into one that is almost practical. Unfortunately, with the additional gain provided by the audio amplifier, we are more likely to hear multiple stations, resulting from the selectivity limitations of our simple tuning circuit discussed in Chapter 2.

Radio listeners of the 1920s were faced with the same selectivity problems, but usually had a lot fewer stations to deal with. After all, if there's only one station within range, you don't have a problem! The answer in the old days was to add vacuum-tube amplifiers with additional tuning circuits to provide more selectivity in front of the detector. These radios were called *tuned radio frequency* sets. While the earliest vacuum tubes were only useable at audio frequencies, it wasn't long before they were developed to the point that they could be used at radio frequencies as well.

A popular arrangement was the "three-dial TRF" receiver. A photograph of an Atwater-Kent three dialer is shown below in **Fig 3-5**. To tune stations the user had to manually adjust each of the three dials to the proper position on the dial. This was better suited for tuning in known stations than searching the dial for a new station. The dials are not calibrated directly in frequency. Rather, each has a 0 to 100 scale, so a tuning chart had to be checked to find out how to tune in each station. This was not for the faint of heart!

What about a TRF set?

Yes, we could make transistor or integrated-circuit RF amplifiers for a radio that include additional tuned circuits. These can separate signals much better, just as the vacuum tube TRF radios our great grandparents used. At the same time we would find that we could pick up additional signals due to the amplification of the signals from the antenna.

The circuit shown in **Fig 3-6** includes a single RF amplifier stage; however, additional ones can be added as needed. With an RF stage, however, we have added a complication that we didn't have to worry about before. I would guess everyone has at some time heard what happens if a microphone is in front of a loudspeaker connected to the output of the same amplifier. The microphone picks up some of the sound from the speaker, amplifies it to produce a louder version, which then hits the microphone. Before long we have an embarrassing howl filling the room. The same thing can happen at radio frequencies. The name of the effect is *oscillation*. In the next chapter we'll describe how to make it happen, but only when we want it to happen.

Oscillation can occur in an RF stage if there is gain (which we surely want) and the output signal

Fig 3-5—Photo of early Atwater-Kent three-dial TRF receiver (*Thanks to Rick Lindquist, N1RL, for equipment loan*).

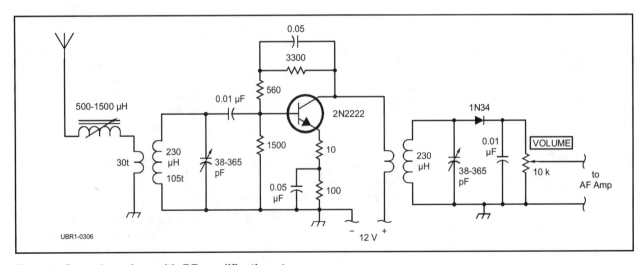

Fig 3-6—Crystal receiver with RF amplification stage.

can couple somehow back to the input of that amplifying stage. This can happen through:
- Coupling within the active device.
- Transformer coupling between the input and output inductors.
- Capacitive coupling between the input and output tuned circuit elements.

We can avoid oscillation if we isolate the input and output circuits through orientation and shielding to minimize any stray coupling. Notice the orientation of the inductors in the old three-dial TRF shown in **Fig 3-7**. Each are at right angles to the other two to minimize transformer coupling. Shielding between stages could also be used to reduce the capacitive coupling between stages. If too much signal couples back through the transistor, we may have to add an additional circuit to cancel out the coupled energy. This is called *neutralization*.

How Far Can You Take This?

You might think that the more RF amplifiers you add, the more gain you can achieve and the shorter the antenna you would need. This is true to a point, but runs out of gas for three reasons:
- The higher the gain at the signal frequency, the more likely oscillation will happen in the RF stages, or the more trouble you have to go through to keep oscillation from occurring.
- Each amplifier adds noise as well as amplifying the signal. The higher the gain, the louder the noise. The resultant noise level detracts from the signal quality. This *internal noise* in the receiver is most evident in the higher HF and VHF ranges, where it can be louder than the *external noise* coming from the antenna.
- External noise due mainly to nearby thunderstorms and powerline noise goes up with lower frequencies. At the MF broadcast band it is relatively high and gets picked up by the antenna with the signal. High amplification ahead of the crystal detector brings up both signal and noise, so it's easier to make signals louder by just turning

Fig 3-7—Inside the Atwater-Kent TRF receiver. Note that the three inductors are perpendicular to each other to reduce unintentional transformer coupling. (*Equpment loan, Rick Lindquist, N1RL*)

up the audio amplifier. Audio amplifiers are generally less complicated and require less special parts than RF amplifiers.

What About That Pesky Crystal?

In addition to amplifying, vacuum tubes also were used as detectors. In the early days of radio, a convenient solid-state semiconductor diode wasn't yet available, although it preceded the transistor (1948) by some years and was in widespread use as a radar detector during WW II. The original crystal detector, with its fussy-to-adjust *cat's whisker*, was a major stumbling block in the way of popular acceptance of radio. Early vacuum-tube detectors, combined with RF and audio amplification, made radio come of age.

The first vacuum-tube detectors were diodes, which served the same function as the fussy cat's-whisker crystal detector. Development using a triode tube led to a number of circuits in which a triode could both detect and amplify, thus replacing a separate first audio amplifier stage.

Where did the TRF go Next?

As noted earlier, a major problem with the TRF was that every added stage added an additional tuning knob! The TRF didn't become popular for casual listening until someone thought of *ganging* the capacitors together through the use

of multisection capacitors turned by a single shaft.

The TRF with ganged capacitors and high-fidelity audio systems moved radio from a hobbyist curiosity to an entertainment and information system given a prominent place in the front parlor. The radio now had just two knobs, a tuning knob and a volume knob. People who just wanted to listen to the radio—not play with it—could now easily use it. The new TRF sets were often in large console cabinets with 12- or 15-inch speakers designed to fill a room with music and news, and some sounded very good.

So why aren't we still using the TRF?

The TRF had two remaining problems. The most serious is that the *bandwidth*, also known as the *selectivity*, was generally a fraction of the tuned-circuit frequency. Thus, if the percentage were 3% using multiple tuned circuits, at the low end of the tuning range we would hear a band of frequencies about 15 kHz wide. But at the other end of the range it would be 45 kHz wide. The usual FCC broadcast-channel assignment is 10 kHz, so obviously with this receiver we would have a problem at the high end of the broadcast band in congested areas or when nighttime ionospheric propagation brings in high-power stations from all over the country.

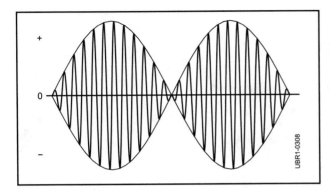

Fig 3-8—Signal envelope input to a detector.

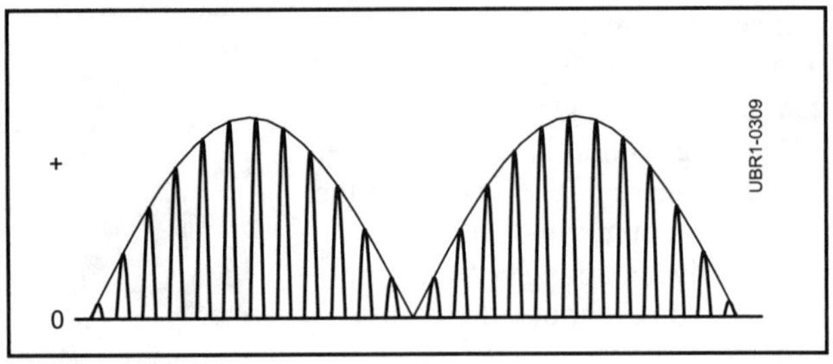

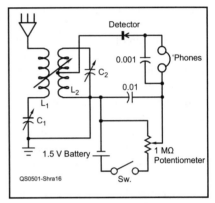

Fig 3-9—Signal output from an ideal detector.

Fig 3-10—Schematic diagram of a crystal receiver using a modern diode with biasing circuit.

The other problem was largely a mechanical issue. The tuned circuits had to be constructed in a more precise way so that all the circuits were on the same frequency at any point of the dial. This was a problem that the three-dial radio didn't have. To compensate for manufacturing irregularities, usually both the inductor and capacitors had *trimming* adjustments. This required an additional manufacturing step of alignment, in which a skilled technician would use calibrated signal generators to adjust all the trimmers so that the radio actually received the frequency shown on the dial and all the circuits were tuned to the same frequency at the same position of the dial. A receiver that did so was said to *track* properly.

TRF receivers are occasionally seen today in special applications, particularly in the case of a receiver that only requires reception of a single frequency.

Modern Diodes—There's Always a Catch!

The use of modern diodes instead of that "chunk of galena and cat's whisker" that granddad fought with may seem like a real blessing. It certainly makes for an easier to build and operate crystal-set radio receiver, but it seems like there's always a gotcha!

To recover the envelope of the transmitted signal, the crystal or diode detector must conduct when the incoming signal is positive and not conduct when the signal is negative. An ideal diode would act like a small resistor to positive-going

signals and look like an insulator to negative ones. **Fig 3-8** shows a signal coming into a detector. **Fig 3-9** shows the output of an ideal diode detector. This is all well and good, but in order to have an ideal detector we need an ideal diode.

With early cat's whisker detectors, it was possible to find a "hot spot" on the material that was the most sensitive. There might have been a number of parameters involved in this hot spot, including the diode forward-conducting voltage.

An ideal diode starts to conduct as soon as the applied voltage becomes greater than zero. Unfortunately, a real diode doesn't start until some forward voltage has been exceeded. For the case of early germanium diodes, about 0.1 V was needed before conduction started. Germanium diodes are less commonly found nowadays, and the more-common silicon diodes require 0.6 V before they start to conduct.

What this means is that the input signal must be much stronger than the diode forward voltage. In the case of a silicon diode, and to a lesser extent early germanium ones, this requires a mighty big signal from a transmitter pretty close by, a very large antenna or both.

What Do We Do Now?

We could go back to galena and cat's whiskers, although galena is no longer readily available at radio dealers. (In fact, radio dealers have become something of a rare species.) An alternative is to *forward bias* the detector. If we apply a dc voltage across the diode, in a way that

doesn't interfere with its operation so that it is *almost* ready to conduct, we can make this problem go away. For example, if we have a diode with a forward turn-on voltage of 0.6 V and we provide a dc bias of 0.599 V, it will start conducting with a signal of 0.001 V, a much easier to get signal from a reasonable antenna than if we needed greater than 0.6 V!

Fig 3-10 shows an example of such a circuit. The 1-MΩ potentiometer can be adjusted to put any voltage between 0 and 1.5 V across the diode. The potentiometer resistance reduces the load on the circuit so that it has minimal impact on detector performance. Note that if the potentiometer is set just below the point at which diode conduction occurs, no current will flow. Even when the diode conducts due to incoming signals, there is no battery current flowing and thus we still do not use any power except that provided by the incoming signal. The battery voltage just moves the free electrons and holes at the diode junction closer together so it's easier for conduction to occur.

The Final Step—the Superhet

Superhet is short for *superheterodyne*. So what's a heterodyne? I'm glad you asked! A heterodyne is a signal produced by two other signals, in a process similar to modulation that we discussed in Chapter 2. The superheterodyne radio receiver uses that method to translate the incoming signal, along with all its information content, to a new frequency called an *intermediate frequency* (IF). The trick is that any frequency we wish to tune to is translated to the same fixed IF frequency for processing. To understand how this happens, we need to introduce a few concepts:

- *Oscillator* circuits to generate new signals
- *Mixer* circuits to perform signal multiplication
- *Filtering* to set the desired bandwidth.

These subjects are basic to radio and electronics. Fortunately they are all straightforward topics based on tools we have used before. We will cover them all in the next few chapters.

Notes

[1] A. Cole, "Practical Pointers on the Audion", *QST*, Mar 1916, pp 41-44.

[2] Wolfgang, L, *Understanding Basic Electronics*, ARRL, Chap 30, p 30-5. Available from the ARRL Bookstore for $20 plus shipping. Order number 3983. Telephone toll-free in the US 888-277-5289, or 860-594-0355; **www.arrl.org/shop/pubsales@arrl.org**.

[3] Ibid, Chap 27, p 27-8, Figure 4.

Review Questions

1. Describe why an audio amplifier makes a crystal set a more practical receiver.
2. How would you be able to tell if a TRF RF amplifier stage were oscillating?
3. Without changing the circuit, how could you make an RF stage stop oscillating?
4. What applications might make use of a single-frequency TRF receiver?

Chapter 4

Oscillators—The Beating Hearts of Modern Radios

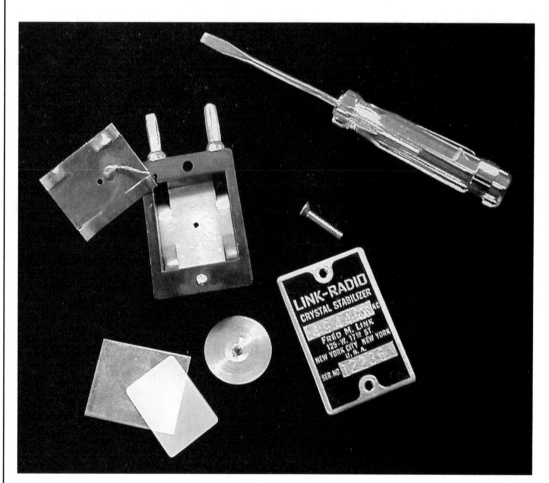

A common WWII type of crystal shown disassembled. The quartz wafer itself is at bottom-center.

Contents

The first function in our quest for the superheterodyne receiver is a signal source. We will begin with the *oscillator*. Oscillators are not only an important part of the superheterodyne receiver; they play an important part in almost all electronic circuits and systems.

As mentioned earlier in Chapter 3, oscillators generate a signal in much the same way a public-address system can generate a signal if the microphone is placed in front of the speaker. As the gain is turned up, a point is reached at which the *feedback* from output to input is amplified enough so that the system breaks into *oscillation* and a large howl (the new signal) results. That's not the result we want in a public-address system, of course, but the effect is one of which we can make good use in a more controlled way.

Feedback Amplifiers

Feedback results from applying a sample of the output of an amplifier (for example) back to its input. Note that we can either apply the sample in-phase with the input signal, which tends to increase the output; or we can apply it out-of-phase to reduce the output. The former is known as *positive feedback*, while the latter is known as *negative feedback*. The two options are shown in **Fig 4-1** for an integrated-circuit operational amplifier (op amp).

To make a positive feedback amplifier into an oscillator we just have to do two things. We can either restrict the feedback signal to the desired frequency of oscillation, or we can amplify only the desired frequency and then make the feedback sufficient so that the amplifier gain times the feedback ratio is greater than one. In either case the feedback will increase until the circuit oscillates at the desired frequency. We will then have an oscillator we can use in a superheterodyne receiver or for many other applications.

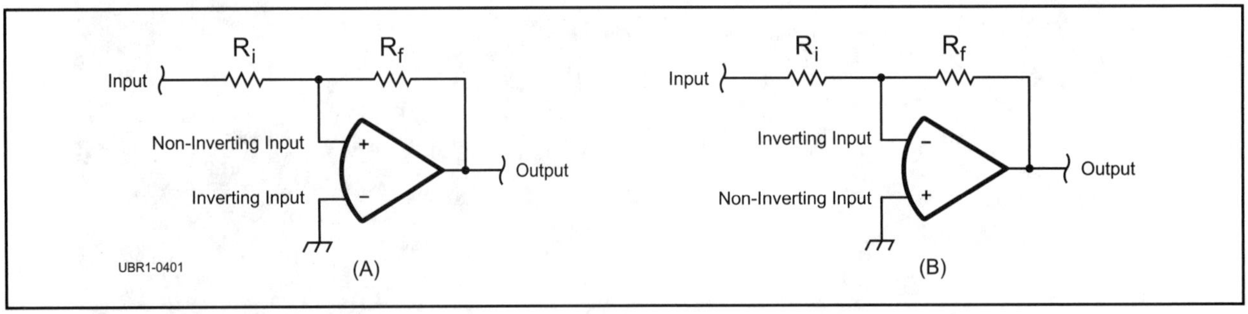

Fig 4-1—Amplifiers with positive (A) and negative (B) feedback.

Types of Oscillators?

Inductance-Capacitance (LC) Oscillators

As noted above, if we want to turn an amplifier into an oscillator at a particular frequency, we must either control the frequency that gets amplified, or control the frequency that gets fed back, or both. **Fig 4-2** shows the RF-amplifier circuit we used ahead of the crystal detector in the TRF receiver described in Chapter 3 (Fig 3-6). In that application, we discussed ways to keep the amplifier from oscillating, since we wanted to hear radio stations—not our own amplifier!

To make the amplifier into an oscillator, all we have to do is put a large-enough sample of the output signal back to the input circuit and tune both circuits to the same frequency. One method of doing this is shown in **Fig 4-3**.

You might wonder what causes the oscillation to start, since we would seem to have no input to begin with. Remember that all amplifiers generate some noise, along with the signals we want them to amplify. That noise generally covers the whole frequency spectrum, including the frequency we are tuned to. Some of the noise on our tuned frequency gets coupled back to the input through the feedback loop. Again, depending on the direction of the connections, this can either add to or subtract from the input signal. In an oscillator we want to add, so we connect the wires in the right direction.

The new signal goes around and around the loop, getting stronger each time, until it reaches the maximum level the circuit can handle. From the output winding we get a sinusoid at the frequency to which the circuits are tuned. We have generated a signal that can be used in a heterodyne receiver (Chapter 9) or as the beginning of a transmitter (Chapter 12). The same oscillator could be used to generate a timing

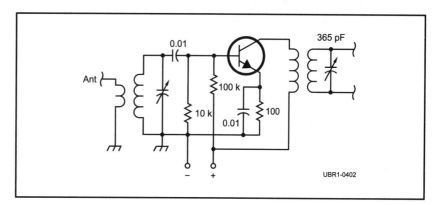

Fig 4-2—RF amplifier without feedback.

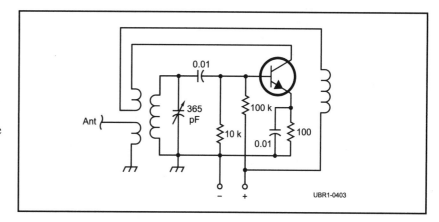

Fig 4-3—RF amplifier with inductively coupled feedback.

reference in a computer, or even in an electronic clock.

Even Better LC Oscillators

The oscillator circuit we just described will work. Sometimes, as noted in Chapter 3, it will even work when we don't want it to! It does have a few areas we can improve. First, it requires two tuned circuits to operate. It would be nice if we could change the frequency with just one knob. Second, the amount of feedback is difficult to precisely set, since it depends on the tightness of coupling between the windings and the location of each coil.

Over the years a number of inventors developed circuits for vacuum-tube oscillators that avoided

these problems. Each still carries the inventor's name, even though the circuits have evolved through four generations of vacuum tubes and a few solid-state devices. The circuits in **Figs 4-4** and **4-5** come from *The ARRL Handbook*[1] and they make use of field-effect transistors (FETs). However, the same tuning and feedback mechanisms can be used with bipolar transistors, integrated circuit amplifiers or vacuum tubes.

The Colpitts oscillator shown in Fig 4-4 neatly solves both problems. Note that the frequency is set by a single tuned circuit composed mostly of L and C1 (I say *mostly* to point out that the combination of C3, C4 and C2 in series are in parallel with C1, limiting the minimum capaci-

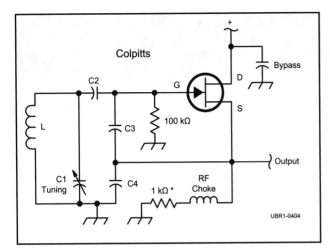

Fig 4-4—The Colpitts oscillator.

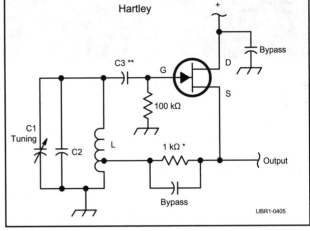

Fig 4-5—The Hartley oscillator.

tance of the circuit and thus the high-frequency end of the tuning range). The feedback is derived from the FET source (at the bottom in the diagram) connection, also the output point. The amount of feedback is set by the impedance of C4 as a fraction of the total of C2 + C3 + C4. The larger the fraction, the more energy gets coupled back into the input circuit. The general rule is to use just enough feedback to allow oscillation, leaving a good signal for the output where we need it.

The Hartley oscillator in Fig 4-5 is very similar in operation to the Colpitts in Fig 4-4, except that the feedback is set by the ratio of the tap point on the inductor L as a fraction of the value of total L. The choice between them may come down to

mechanical issues. For example, in many applications the *stability* (the ability of an oscillator to stay on frequency in spite of environmental changes) is important. Often movement of the coil due to shock and vibration is a problem and more connections may make secure mounting more difficult.

Another consideration is frequency range. If a large frequency range is needed, sometimes different values of L are switched in using multiple inductors selected with a band switch. The Colpitts only requires switching two wires; the Hartley requires three. The Hartley has the advantage on total parts count, however.

There are variations of these circuits, both in terms of active

device and exact tuning methods, but these are the two most common LC oscillator circuits. They are especially useful ways to have an oscillator that is mechanically tuned. By the way, a variable capacitor is shown as the tuning element in the LC circuit; however, the frequency can be changed as well using a fixed capacitor and a variable inductor. Some of the highest quality variable frequency oscillators (VFOs) over the years have used a precision threaded rod to move an iron core in and out of an inductor to vary the frequency.

To remember the names of the circuits, note that Colpitts uses feedback through capacitors ("C"), while the Hartley uses inductors (units of "H").

Crystal Oscillators

A completely different type of crystal than the one used in a crystal radio in Chapter 2 is the *piezoelectric crystal*. This kind of crystal is made from quartz crystals, just as you may have seen growing in rock formations. They have an interesting and useful property. If a voltage is applied across them, an electric field results and this causes a mechanical motion of the structure. When the voltage is removed, the crystal returns to its original configuration. Just like pushing a mechanical weight suspended from a spring, the crystal has a natural frequency at which it wishes to sway to and fro. The frequency of oscillation depends on the dimensions of the crystal. Representative crystal types are shown in **Fig 4-6**.

By precise cutting, grinding and polishing, quartz crystals can be made to oscillate at a specific frequency. Functionally, they operate as if they were an LC resonant circuit. Such a crystal can be used almost anywhere an LC circuit can be used, including acting as the frequency control for an oscillator. **Fig 4-7A** shows a crystal oscillator in a Colpitts configuration. Note that in a crystal oscillator we don't have a coil tap across an equivalent resonating inductance, so we can't make a Hartley circuit.

The 30-pF variable capacitor in Fig 4-7A can be used to slightly shift

Fig 4-6—Some examples of frequency-control crystals. The types in the metal cans are sealed, while earlier types dating from WW II can be opened to expose the parts.

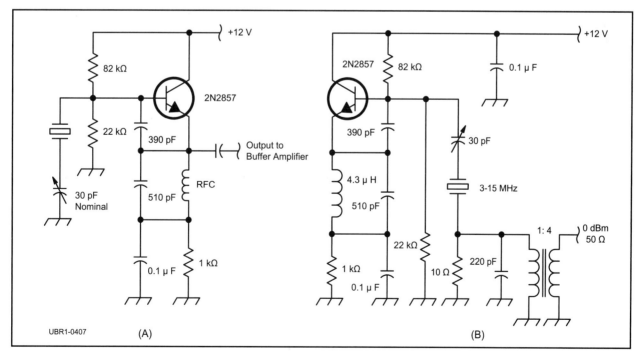

UBR1-0407   (A)   (B)

Fig 4-7—Crystal-controlled Colpitts oscillators.

the oscillation frequency. The crystal must be specifically cut taking into account the circuit configuration and capacitance into which it operates.

While the oscillator in Fig 4-7A uses the crystal as a part of the equivalent amplifier tuning network, it is also possible to use the crystal in the feedback and output path to allow only the desired frequency to operate. There are a number of such configurations and Fig 4-7B shows one.

Crystal oscillators have one major advantage over LC oscillators—they are stable as long as the size of the crystal itself remains constant. Of course, this will change slightly with temperature, and thus crystals used in oscillators requiring high precision are often mounted in a temperature-controlled miniature oven designed to avoid frequency changes from temperature shifts.

One disadvantage of crystals is that they will operate only on a single frequency. In many applications, this can be overcome by having multiple crystals and changing to the desired frequency by switching crystals.

In addition to the quartz crystals we mentioned, there are some salt crystals, such as those made of Rochelle salts, that have useful properties in the audio frequency range. When driven by an audio signal, they can be used as headphones or even as small speakers. Such crystals also work the other way. If you stress them and release them they will generate a voltage waveform related to the motion. This makes them useful as the element in a microphone.

Notes

[1] *The ARRL Handbook for Radio Communication*, 2005 Ed. See Chapter 10 for detailed technical information on oscillators and synthesizers. Available from the ARRL Bookstore for $39.95 plus $8 shipping in the US, $10 elsewhere. Order number 9280. Telephone toll-free in the US 888-277-5289, 860-594-0355; or **www.arrl.org/shop/ pubsales@arrl.org**.

Review Questions

1. Describe some of the limitations of LC oscillators intended for precision or high-stability applications. Describe alternatives for fixed and variable-frequency applications. What problems do they solve?

2. A crystal manufacturer provides crystals with a frequency tolerance of ±.005%. If you purchase such a crystal for 7.0 MHz, what range of frequencies might you end up with?

3. You need to provide a reference oscillator to generate a frequency of 10.125 MHz. Your requirements are that you must be close enough to the design frequency that your signal be between 10.1245 and 10.1255 MHz. What frequency tolerance do you need to provide?

Chapter 5

Frequency Synthesizers

UR3IQO's homebrew fractional-N synthesizer from Nov/Dec 2003 *QEX*.

Contents

The ideal oscillator has long been considered a combination of the LC oscillator, with its wide adjustment range, and the crystal oscillator, with its precision and stability. Over the years there have been many efforts to make an oscillator that provided the benefits of both. Some have been more successful than others. Early efforts used multiple crystal oscillators mixed together, as will be described in the next chapter. However, modern digital integrated circuits have made this task easier and less expensive. We will discuss the two most frequently encountered methods of frequency synthesis.

But first we will discuss electrically tuned oscillators, a key element in many synthesizers.

Electrically Tuned Oscillators

The variable oscillators we have talked about so far have used mechanical rotation of a variable capacitor or an inductor's tuning slug to change frequency. Often we want to change an oscillator frequency in response to an electrical command, either for frequency change from a remote position or sometimes to correct for frequency errors.

In the early days of radio, this was accomplished by attaching remotely controlled electric motors, called *servomotors*, to a capacitor shaft and having the motor turn the shaft as if it were a local operator.

The semiconductor revolution has brought us a more elegant way to accomplish this task, and without moving parts! We have discussed the use of semiconductor diodes as receiver detectors in Chapter 2. There, we were most concerned with diodes when they were conducting, but here we will focus on when they're not conducting.

In a diode connected so it does not conduct, as shown in **Fig 5-1**, a *depletion region* exists between the N and P material. It is just as if there were two pieces of metal with an insulator (the depletion region) in-between. Well, we could also call that a capacitor! The interesting property of this configuration is that as the voltage is increased, the free

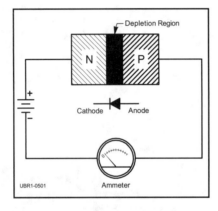

Fig 5-1—A semiconductor diode connected so no current flows. The depletion region in the middle acts as an insulator. With two conductors sandwiching an insulator in-between, this looks exactly like a capacitor!

electrons in the N material and holes in the P material are attracted closer to the diode terminals. This makes the insulator (depletion region) wider and decreases the capacitance of the diode. This is the effect we want— We can vary the capacitance of this device just by changing the voltage. No moving parts, and it doesn't draw any current so no power is consumed. It's almost free.

Semiconductor diodes are made especially for this purpose and are called *varactors* or tuning diodes. An example of a Colpitts oscillator with voltage tuning is shown in **Fig 5-2**. Note that the same arrangement can be used to change the frequency of a crystal oscillator, although the change is generally less because the main thing that determines the frequency is the crystal itself.

The Phase-Locked-Loop

The phase-locked-loop (PLL) combines digital circuitry with an analog voltage-controlled oscillator (VCO), as described above. The system causes the VCO to be adjusted in steps that are as accurate

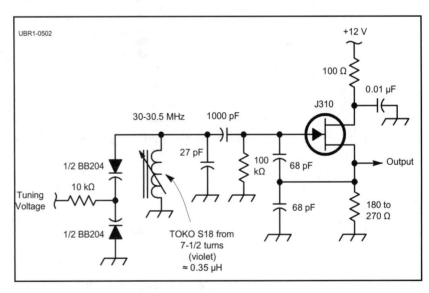

Fig 5-2—A voltage-controlled Colpitts oscillator.

as a crystal, but covers the whole frequency range of the VCO. The configuration is shown in **Fig 5-3**.

The basic operation is straightforward. But as usual, the devil is in the details. A voltage-controlled oscillator (VCO) covers the range of frequencies desired. If left to its own devices, it would just be an LC controlled oscillator—convenient, but not very stable. To synthesize the stability of a crystal, we also need a *reference* crystal oscillator. We select the desired frequency by using digital dividers to divide down the frequencies of both the reference oscillator and the VCO (by different numbers) until they are the same. We then compare the two using a *phase detector* (essentially a mixer, to be described later) and use any difference to drive the VCO until the input frequencies to the phase detector are the same.

This system is called a *phase-locked loop* and is designed to operate quickly as soon as there is any phase difference between the divided oscillator and the divided reference. Earlier implementations of such systems were based on a frequency-locked loop instead of a phase-locked loop. However, by its nature a frequency-locked loop takes multiple cycles to determine frequency error, while theoretically a

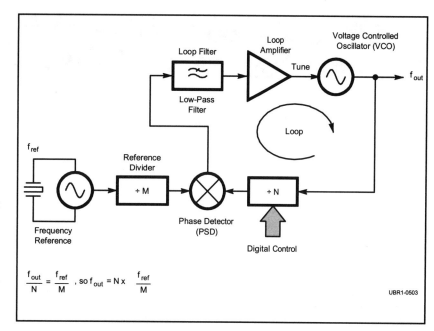

Fig 5-3—A phase-locked-loop controlled stable oscillator.

phase-locked loop takes less than a cycle to measure and react to a phase difference. In practice, the speed may come out to about the same, since all loops contain smoothing filters to keep the loops from responding so quickly that they *hunt* back and forth trying to "lock up."

Early frequency synthesizers were sometimes called *crystal synthesizers*. This conveyed the idea that the output frequency would be as stable as the

crystal reference.

Synthesizers are now found in most modern variable-frequency equipment. Like all apparent good things, however, synthesizers do have limitations. The most serious may be that the signal often contains extraneous components resulting from the correction signal. For the purest of signals, traditional LC or crystal oscillator circuits are still employed.

More Frequency Synthesizers

Direct Digital Synthesis (DDS)

The frequency synthesizers just described bring together the technologies of analog oscillator circuits and dedicated digital circuits performing a control function. The actual signal still comes from an LC oscillator, however.

Direct Digital Synthesis (DDS) takes the migration to digital circuitry the final step. In a DDS signal source, there is no oscillator dedicated to this function. A microcomputer generates the signal instead, completely eliminating a direct connection to an analog circuit.

Even so, somewhere in the microprocessor architecture there is a *clock* function that serves as a time base for all processor activity and in reality is the standard against which the DDS process happens. What's different here is that this clock crystal oscillator is further removed from the signal generation process, and it supports all processor timekeeping functions. After all is said and done, however, the accuracy of the process can be no better than the reference crystal.

So How's it Work?

Computers, especially *sound cards*, contain devices known as digital-to-analog converters or *DACs*. A DAC reads in a binary based number, let's say 01010101, decodes it as 85 (decimal) and puts out an analog voltage related to the decimal value—let's say 85/100 or 0.85 volts. If we can select any number, we can define the voltages corresponding to a sine wave. Let's say we want to represent one cycle of a sine wave by 10 values. We can easily calculate them as follows:

Angle	Sine of Angle
0°	0
36°	0.588
72°	0.951
108°	0.951
144°	0.588
180°	0
216°	-0.588
252°	-0.951
288°	-0.951
324°	-0.588

Since computer memory is cheap, we could store these values in memory. In a dedicated memory device, we could put them in a read-only-memory (ROM) and forever have values associated with an almost perfect sine wave, in as many increments as we want. We could just as easily have 1000 points as 10.

Now we put the computer to work. We merely write code that retrieves the values from ROM in sequence and applies them to the DAC to form our analog output. We filter to smooth out between the samples, just as we did the audio output of the crystal set receiver, and we have generated very close to a sine wave. If we have 1000 points, the output frequency will just be equal to the number of times we go around all the memory locations in a second divided by 1000. To set the frequency, we just program the computer to do it at the correct rate. With modern computers being clocked well above 1000 MHz, we can make this happen at pretty high frequencies without working the processor too hard.

Hybrid Synthesizers

The DDS synthesizer offers high resolution (a small step size), while a PLL synthesizer offers a wide frequency range. DSS synthesizers are available as dedicated processors, all on a single chip to achieve reasonable costs. Many synthesizer designs combine the two technologies, to get the benefits of both.

Fig 5-4 is a greatly simplified version of such a design. A DDS is used to generate a sine wave with a

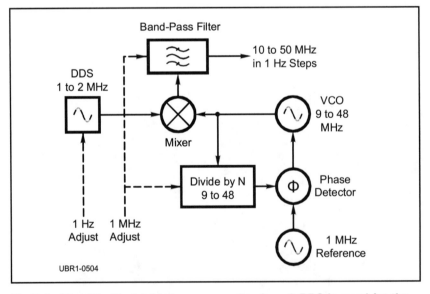

Fig 5-4—Simplified hybrid frequency synthesizer. A DDS is used for the small increments in frequency, while a PLL is used for larger frequency steps.

frequency from 1 to 2 MHz, in 1-Hz steps. A PLL is used to generate a signal from 9 to 48 MHz, in 1-MHz steps. A mixer (which we'll cover in Chapter 6) combines the signals and takes the sum of the two. A filter is used to make sure the correct frequency is selected.

If we wanted to generate a frequency of, let's say, 12.5673 MHz, we would have the DDS in Fig5-4 generate a 1.5673-MHz signal. We would set the ÷ N divider to divide by 11 and set up the bandpass output filter after the mixer to pass from 12 to 13 MHz. The mixer would sum the 11 MHz from the PLL and 1.5673 to get 12.5673 MHz. The filter is used to eliminate undesired mixer products. And we thus have generated our desired signal, ready for use.

Review Questions

1. Describe advantages of a synthesized frequency generator as compared to crystal oscillators. What are the disadvantages?

2. Repeat question 1 in comparison with tunable L-C oscillators.

3. What special requirements are placed on the power supply of synthesizers using voltage controlled oscillators as shown in Fig 5-2?

Chapter 6

Mixing It Up With A Mixer

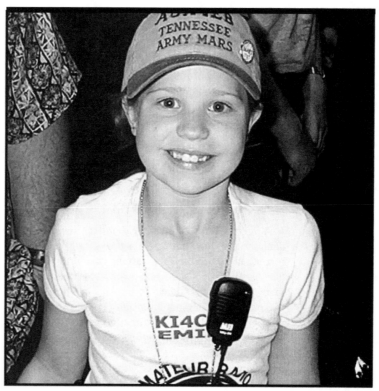

Who says youngsters are rare in ham radio? Nine year old ARRL Member Emily (Chick) Bishop from Cleveland, TN, checks in at the Dayton 2004 ARRL booth with her dad, Mike, KI4AME.

Contents

In Chapter 1 we briefly discussed multiplying two signals to modulate the amplitude of one by the other. Fig 2-7 in Chapter 2 showed the graphical result of the multiplication of two sine waves plotted against time. As you may remember from high school trigonometry, the result of multiplying two sinusoids is well defined by the following trigonometric identity:

$$\sin A \, \sin B = \frac{1}{2}[\cos(A - B) - \cos(A + B)]$$

Eq 6-1

This is a very important result and we will make use of it in a number of ways. Note what it tells us. If we multiply sine waves at two frequencies, the results consist of sine waves at two new frequencies, the sum and difference of the two original signals. The fact that they are cosines rather than sines just means that they are shifted 90° from the original signals. The scale factor of $\frac{1}{2}$ just means that each of the new signals is half the amplitude of the originals. Note also, in this idealized multiplication, the original signals are no longer there at all. We not only made new signals, we made the old ones disappear!

For the earlier case of a 600 kHz radio signal multiplied by a 600 Hz audio tone, Eq 6-1 says we will have two radio signals, one at (600,000 − 600) Hz, or 599.4 kHz, and the other at (600,000 + 600) Hz, or 600.6 kHz. In a regular AM system, we don't cancel out the carrier frequency so we would also have the original 600 kHz carrier signal. The result is that we would have three signals as shown in **Fig 6-1**, a plot of amplitude versus frequency.

This representation is also called a *frequency domain* plot, in contrast to Fig 2-7, which is called a *time domain* plot. Both are representations of the same signals. Note particularly that this result doesn't say anything about one signal being a radio signal and one an audio signal. It applies to *any* two signals.

How we get Multiplication

A *mixer* circuit is the place that the actual multiplication takes place. A mixer can be used in many applications, and as a consequence has many names—first detector, translator, modulator, demodulator and multiplier. In all cases the function is simple—to perform the multiplication of two signals. Let us look into this multiplication business some

more, using a linear amplifier for illustration.

When we build an amplifier, a goal is to have at the output an exact copy (except larger) of the input signal. If the amplifier gain is K and the input signal is x, we generally look for an output y = K × x. This is known as a *linear system* since the relationship between y and x is a straight line with a slope of K. The output versus input (known as the *transfer function*) of a linear amplifier is shown in **Fig 6-2**.

The true test of linearity is to apply two signals simultaneously and make sure that the only outputs are replicas of the two inputs. Expressed mathematically: If the two signals are a and b, the output becomes, y = K × (a + b) = K × a + K × b. If a and b are sinusoids at two frequencies, we just get two louder sinusoids at the output. In such an amplifier, this is exactly what we want. If we are listening to a violin and an oboe, we don't want the amplifier to generate the sounds of a piano and flute! This ideal amplifier will not serve as a mixer.

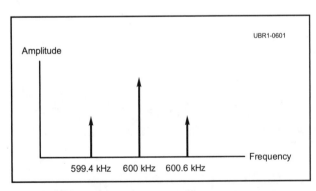

Fig 6-1—Frequency-domain representation of multiplication of the two signals with a carrier.

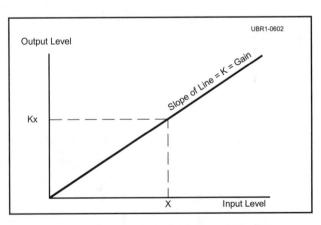

Fig 6-2—Input-output relationship of an ideal linear amplifier.

The diode has a more complicated transfer function, shown in simplified form below in **Fig 6-3**. Note that if the diode is operated well above the forward-biased conduction point, it is quite linear and looks a lot like a resistor. If we operate it well below the conduction point, it is also linear, but it looks like an insulator (y = 0 × x). For a narrow range of x just above the point at which current starts to flow, it has a response that looks a bit like a parabola, with a response close to $y = K \times x^2$.

In this case, if we apply two signals, we get a more complicated result. For inputs a and b, we have:

$$y = K \times (a + b)^2 = K \times (a^2 + 2ab + b^2)$$

$$\text{(Eq 6-2)}$$

Note what we have here—the middle term represents the product of the two inputs, just what we wanted! We also have the square of each input, resulting in a twice-frequency term. But these extraneous results will not be a problem if we design our system carefully. Taking the case of our earlier example of a modulator, for example, with 600 Hz and 600 kHz signals, we have unwanted signals at 1200 Hz and 1200 kHz, both easy to eliminate with simple filtering from our desired 600 kHz ± 600 Hz signals.

Frequency Translation

While moving around a single 600 Hz signal up to region around 600 kHz may be of academic interest, most practical applications involve *bands* of frequencies. Telephone-quality speech covers audio frequencies from 300 to 3300 Hz. (This is called "toll quality"—as in making a "toll call," for which the telephone company charges you.) AM broadcast-quality audio extends from around 20 to 5000 Hz, while hi-fi music is generally considered to be within the range of 20 to 20,000 Hz.

When we modulate a transmitter, we usually use signals within one of the above ranges, depending on the kind of information we wish to send. Let's take our 600-kHz carrier as an example. Since it's in the AM broadcast band, we will assume the information is contained in the 20 to 5000 Hz range. A simple transmitter might consist of an audio source (from the announcer's microphone for example), a carrier signal at the assigned frequency of 600 kHz and a mixer (also known as a *modulator* in this application) to bring them all together. This is shown in block diagram form in **Fig 6-4**.

Note that we have effectively translated the frequency range of the modulating, or information, signal up to a frequency at which it can be sent via radio. The resulting spectrum is shown in **Fig 6-5**.

Mixer Circuits

The simplest mixer circuit is our earlier diode crystal set. When used as a receiver of AM signals the *detection process* can also be described mathematically as a multiplication process. If we receive a signal for a 600-kHz carrier, amplitude modulated by a single tone of 600 Hz, we have shown that

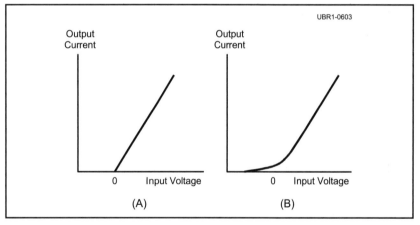

Fig 6-3—At A, an ideal diode response compared to a real diode's response at B.

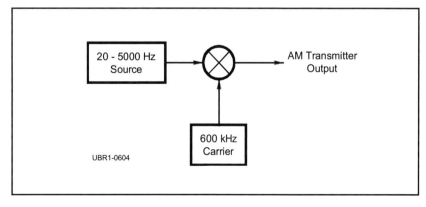

Fig 6-4—Simple AM radio transmitter.

the signal is represented by the carrier surrounded by frequencies of 599.4 and 600.6 kHz. All three frequencies transmitted are in Fig 6-1 at the input of the diode. Since there are products of equal amplitude and phase on either side of the 600-kHz carrier they add up at the output of the diode detector together, eliminating the factor of $^1/_2$ in Eq 6-1. At the output of the diode, the twice-carrier frequency terms are all around 1200 kHz, well above the response of our headphones, audio amplifier or ears.

The single-diode mixer is sometimes encountered in actual practice, often in simple microwave receivers. For most other applications, somewhat more complicated configurations are usually employed. Often they are variants of the single diode detector, using the multiple diode ring or balanced mixer. A circuit for such scheme is shown in **Fig 6-6**.

This mixer has a number of advantages over the single diode mixer. A major one is that the input signals cancel at the output, so they are eliminated from the process.

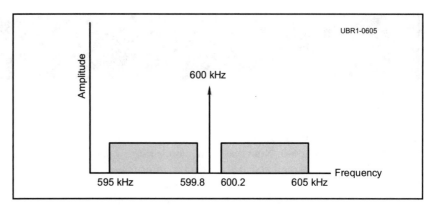

Fig 6-5—Frequency-domain representation of the output of simple AM radio transmitter.

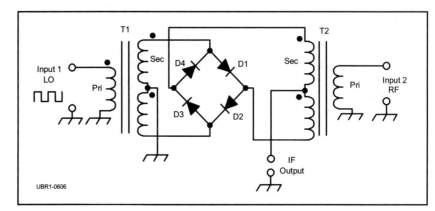

Fig 6-6—Diode-ring balanced mixer.

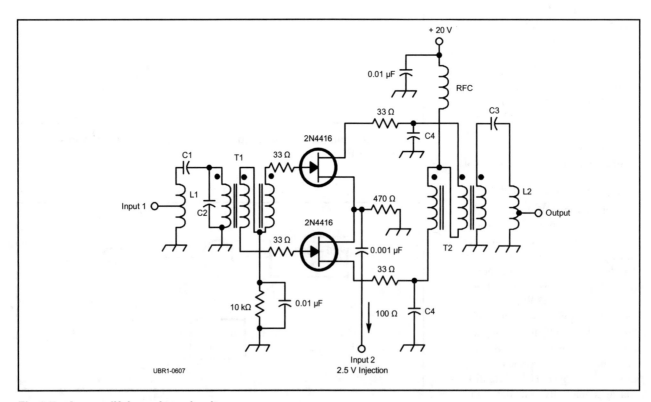

Fig 6-7—An amplifying mixer circuit.

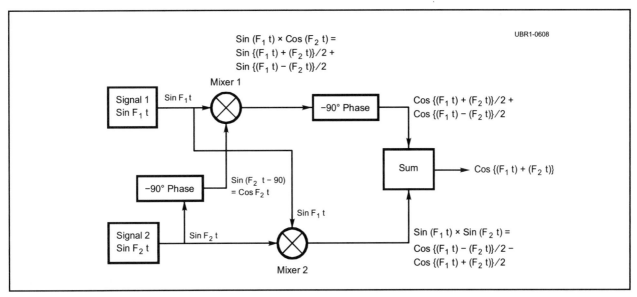

Fig 6-8—Image rejecting mixer, block diagram and signal relationships.

Another advantage is that the coupling transformers for each signal are separate and thus can be optimized for each frequency range.

None of the mixer circuits described so far provide any gain. The result is an output signal less than any of the inputs. For this type of mixer, any increase in signal level, if needed, must occur in an amplifier following the mixer. It is also possible to have amplifying mixer circuits that take advantage of an active amplifier circuit with a non-linear input range, or a process in which one signal multiplies the other by changing the effective gain. A representative amplifying mixer circuits is shown in **Fig 6-7**.

Mixer Image Response

One fact to keep in mind about all the mixers we have discussed so far is that they provide signals at both the sum and the difference frequencies of the two inputs. Sometimes, as in an AM modulated transmitter, that's exactly what we want. In most cases, we would really rather have just one output frequency. In many mixers, it's actually worse than that because we also get harmonics of all

the input signals and combinations of these harmonics. A truly *ideal* mixer is hard to find! Generally, we resort to filtering (discussed later in Chapter 7) to eliminate the outputs we don't want, which are often called *images* because they are symmetrical around the carrier frequency.

Image-Canceling Mixer

An interesting if somewhat more complicated mixer is known as an *image-canceling mixer*. The word canceling may be a bit optimistic, as we will discuss, so sometimes the word *rejecting* is used in place of

canceling. This clever design makes use of two mixers and two phase-shift networks to cancel either of the outputs, depending on the phase shifts of the two arms. The design is shown in **Fig 6-8** along with the description of the trigonometric relations that result in the desired operation. This design makes use of additional trigonometry relationships similar to the one at the beginning of this chapter. Note that at the summation point on the right of the figure, the difference term cancels out, leaving just the sum of the two input frequencies.

Review Questions

1. If you use a simple mixer to multiply signals of 10 MHz and 1 MHz, what are the frequencies of the resulting signals?

2. If you wish to translate a signal at 14 MHz to 9 MHz for further processing, and you use a signal at 5 MHz for your other signal, signals at what other frequency will be translated to 9 MHz?

3. The image-canceling mixer of Fig 6-8 provides an output of only the sum of F1 and F2. How could it be rearranged to provide the difference instead of the sum?

Chapter 7

Filtering Out What You Don't Want

Photo of author Joel Hallas, W1ZR, at his home station.

Contents

In Chapters 2 and 3, we discussed the use of resonant circuits to separate stations in a simple crystal set. Using such circuits for that purpose is a form of *filtering*. A short description of filtering is to note that a filter passes selected frequencies and reduces or *attenuates* others. There are a number of places within radio and other types of electronic equipment where filtering plays a critical role. This chapter will start to describe some of the types of filters that you will encounter in radio equipment.

Filter Functions

Filters can be separated into categories both by what they do and by how they do it. In terms of what they do the general classifications are:

• *Low-pass filters*. These pass signals with frequencies below a *cut-off frequency* and attenuate, or reduce, signals with frequencies above that cut-off frequency.

• *High-pass filters*. Not surprisingly, these pass signals with frequencies above a cut-off frequency and attenuate signals with frequencies below that cut-off frequency.

• *Band-pass filters*. Band-pass filters pass signals with frequencies within a band of frequencies. If you combine a high-pass and low-pass filter, with the high-pass having a

cut-off frequency below the low-pass cut-off frequency, you will have a band-pass filter.

• *Band-reject filters*. Band-reject filters pass signals except for those with frequencies within a selected band of frequencies. If you combine a high-pass and low-pass filter, with the high-pass having a cut-off frequency above the low-pass cut-off frequency, you will have a band-reject filter.

One-Element Filters

The simplest filters are those with one element. More complicated filters are just combinations of multiple elements, so single element filters make a great place to start. The first filter I discuss is not a filter at all, but should get us ready to start looking at real ones. Let's invent the term *all-pass filter*. Based on the definitions above, you would expect such a filter to pass all frequencies and reject none. It may not be very interesting but it will provide a starting point.

Fig 7-1 is a schematic of such a device. It is composed of two resistors of equal value connected in an "L" configuration. The source is on the left and applies a signal to the two resistors in series. The output is taken across the shunt element or *load resistor*, R_L.

Note that from Ohm's law, we can compute the current in the circuit as

$I = E_S/R$

where R is the series combination of R_1 and R_L.

So

$I = E_S/(R_1 + R_L)$

The output voltage is just

$E = I \times R$

or in our case

$E_0 = [E_S/(R_1 + R_L)] \times R_L$

To find the ratio of output voltage to input voltage, we just recombine our terms and divide each side by E_S to yield

$E_0/E_S = R_L/(R_1 + R_L)$

This relationship is often called a *transfer function*, since it expresses the amount of signal that transfers through the network.

The transfer function for this circuit has no dependence on frequency. So if you plotted the function against frequency you would see a straight line, as shown in **Fig 7-2**.

Suppose we replace resistor R_1 with an inductor? See **Fig 7-3**. What will this do to our transfer function? At low frequencies (we'll work some numbers shortly), the reactance of the inductor will be small compared to the resistor and the transfer function will be almost unity. Above some high frequency, the reactance of the inductor will be much higher

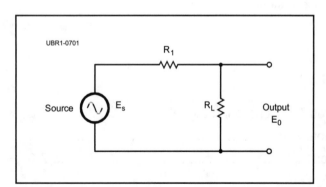

Fig 7-1—Schematic diagram of "all-pass" filter.

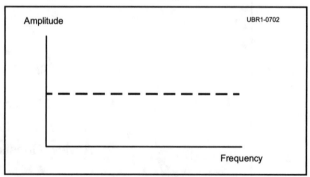

Fig 7-2—Transfer function of all-pass filter.

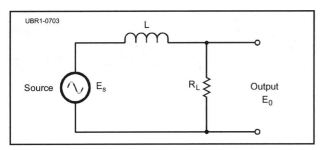

Fig 7-3—Schematic diagram of single-element low-pass filter.

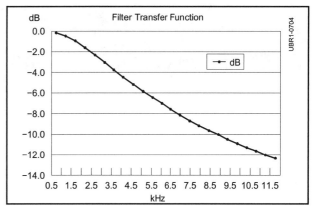

Fig 7-4—Single-element RL filter response.

than the resistance and the transfer function will be almost zero. This sounds *a lot* like a low-pass filter, since the lows get through and the highs don't!

Now let's do some real numbers. Remember that the reactance of an inductor is

$$X_L = 2 \times \pi \times F \times L$$

Let's pick a load resistance of 600 Ω (this will usually depend on where the filtered signal goes next), and decide that we want the reactance to equal the resistance at 3000 Hz. To find the value of L, we solve the equation above for L and get:

$$L = X_L / (2 \times \pi \times F)$$

Since we know all the values, we can substitute and find

$$L = 600 / (2 \times p \times 3000) = 0.032 \text{ H}$$

which is 32 mH.

Now that we've designed a low-pass filter, let's analyze how it will work. As in the case of our all-pass filter, we will first determine the current through the filter as a function of frequency and then

determine the output voltage at the load resistor. Remember that the magnitude of the impedance of an RL circuit is not just the sum of their impedances but is equal to the square root of the sum of squares of the resistance and reactance. So at our design frequency,

$$Z = \sqrt{\left(R^2 + X_L{}^2\right)} = \sqrt{600^2 + 600^2} = 848.5\,\Omega$$

The current magnitude (we could also compute the phase if we wanted to) of the output is just $I = E_S/Z$ and the voltage is just $I \times R_L$. The output voltage can be determined as

$$E_O = (E_S/Z) \times R_L$$

The resulting transfer function is

$$E_O/E_S = R_L/Z$$

We could solve for the output at 3000 Hz and get 600/848.5 or 0.707 E_S. This represents a reduction in signal of 3 dB, or half of the power applied to the input goes to the load. We could solve for the output as a function of frequency.

I show some values of interest in **Fig 7-4**. Note that the output of the

filter is half the input power at the design frequency, where the inductive reactance equals the resistance. At twice that frequency, the output is down by another half power, an additional 3 dB. Doubling the frequency is referred to as an *octave*—this also includes eight musical notes. A filter that has such a response is often referred to as a "3-dB-per-octave" filter.

If we replaced the inductor with a capacitor, we could have a high-pass rather than low-pass filter. The design approach and analysis are similar to the example presented above.

The simple one-element filter in Fig 7-3 is useable, but does not have a sufficiently rapid *roll-off* for many applications. The roll-off is often referred to as the filter's *skirts*, especially in the case of a band-pass filter. This filter would work well to separate audio from radio frequencies, but often the requirements are much more stringent. In the next section, we will discuss how we can achieve better filtering. You guessed it—with more complicated filters!

For some applications, a single-element filter will provide the needed amount of filtering. In most cases, however, the requirement for filtering is more demanding. We have previously talked about some filter applications, such as those used to separate adjacent radio channels. In an ideal AM broadcast receiver, we would have a 10-kHz wide radio channel assigned to one radio station and the next station on the dial could be as close as 10 kHz away. Note that each channel extends 5 kHz either side of the carrier. Thus the high end of one channel is exactly 0 Hz away from the low end of the next. Our ideal receiver filter would accept the full 10-kHz channel that we wanted to listen to and would completely reject the channels above and below it. The response of such a band-pass filter would look like that shown in **Fig 7-5**. An ideal low-pass filter response is shown in **Fig 7-6.**

So How Do We Get There From Here

The general rule is that to get *better* filtering, we need *more* filtering. As with anything else, more is not quite enough—it must be the correct *more* to do the job! One way to apply additional filtering is to just add additional sections to our simple filter. Each time we add an additional identical section, we reduce the output at the cut-off frequency by an additional 3 dB. Of more significance is that the filter skirts get steeper with each added section. Note that if we define our cut-off frequency as the frequency at which we are down 3 dB, we must move each section slightly to accommodate the effect of the multiple sections. For example, if we have three sections, the output at the cut-off frequency should be set at a level of 1 dB in each section so that the total attenuation of the combined filter is 3 dB at the cut-off frequency.

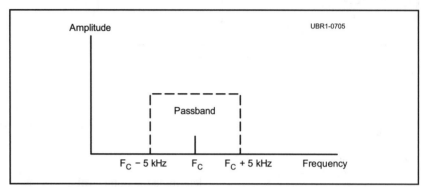

Fig 7-5—Transfer function of an ideal band-pass filter with infinitely steep skirts.

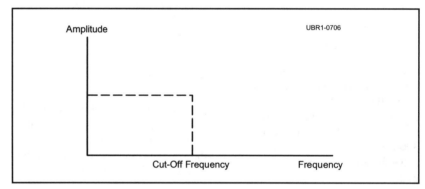

Fig 7-6—Transfer function of an ideal low-pass filter with infinitely steep skirts.

Another way to increase the effectiveness of a filter is to use both inductance *and* capacitance in each section. Our simple one-element filter used a series inductance or capacitance in a low or high pass filter respectively. By eliminating the shunt resistor in the intermediate stages, and using a shunt capacitor instead or the resistor in a low-pass filter, for example, we have twice the effect with frequency as we have with just one reactive element per section. As the frequency goes up, not only does the inductor series reactance increase, but the capacitor shunt reactance decreases, both resulting in a lower voltage at the output of the section.

An example of a low-pass filter

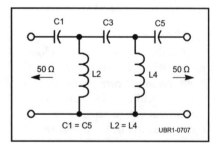

Fig 7-7—Schematic diagram of five element low-pass filter.

with five reactive elements is shown in **Fig 7-7**. Note that we can keep adding elements and adjusting the values and the response will continue to have progressively steeper slopes. The actual values of the elements for

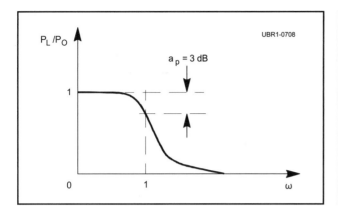

Fig 7-8—Transfer function of a Butterworth low-pass filter with finite skirts.

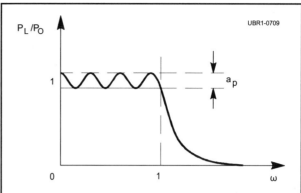

Fig 7-9—Transfer function of a Chebyshev low-pass filter showing passband ripple.

such filters have been studied at great lengths and tables of values are available for optimum filters based on a number of criteria. The two most common formulations are:

- Butterworth—This set of filter values provides the best skirt slope possible while maintaining a flat response in the pass region. An example of the Butterworth response is shown in **Fig 7-8**.
- Chebyshev—This set of filter values provides sharper skirt slopes than the Butterworth, at a cost of some ripple in the pass-band. An example of the Chebyshev response is shown in **Fig 7-9**.

The effect of adding elements is shown graphically in **Fig 7-10**. Here the skirts are shown for 1 to 10 elements in a Butterworth design, with all other parameters adjusted to keep the curves lined up at the edges. The change in slope of the skirts shows up dramatically.

Band-Pass Filters

In many applications, a band-pass response is desired, such as the case where we wish to separate out a single channel from other nearby radio channels. One way to get this result is by combining low and high pass filters, as shown in the examples in **Figs 7-11** to **7-13**. This technique is particularly effective if the desired band edges are separated

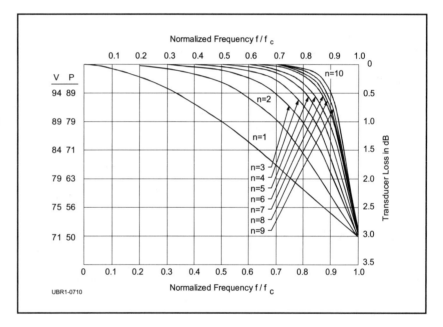

Fig 7-10—Effect of adding elements to the skirts of a Butterworth low-pass filter.

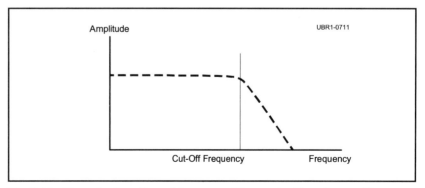

Fig 7-11—Transfer function of low-pass filter part of band-pass filter combination.

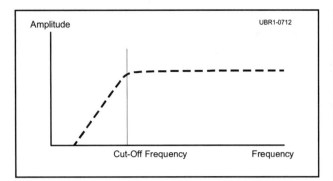

Fig 7-12—Transfer function of a high-pass filter part of band-pass filter combination

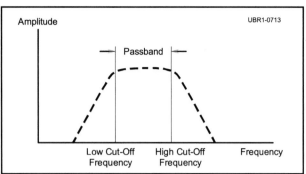

Fig 7-13—Transfer function of resulting band-pass filter.

by a relatively large difference in frequency.

In cases when the desired pass-band is a small fraction of the center frequency, it's better to use a filter that starts off as a band-pass filter. You will remember that we used a *tuned circuit*, a parallel combination of inductor and capacitor, resonant at our desired receive frequency to select a station with the simple crystal set in earlier chapters. Such a tuned circuit forms the basis of many band-pass filters. The impedance of a single parallel tuned circuit is shown in **Fig 7-14**. Note that the sharpness of the filter is a function of the Q. For most lumped-constant circuits, this is the ratio of the reactance to the resistance. In such circuits, the majority of the resistance is in the wire of the inductor, although other losses can act in the same way.

It can be shown that the half-power (3-dB) bandwidth of such a circuit is just the Q divided by the resonant frequency. While a parallel-resonant circuit has a high imped-ance at its resonant frequency, and can thus develop a high voltage across it at resonance, a series-resonant circuit has a low imped-ance. It allows high current to pass at resonance. It would seem that we could use such circuits as the basis of a band-pass filter of any desired relative bandwidth, however, note from Fig 7-14 that as the Q is decreased, the slope of the skirts gets less and less steep, allowing other signals to enter the pass band.

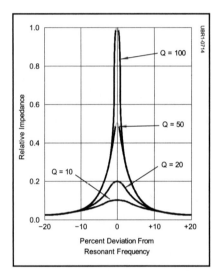

Fig 7-14—Relative impedance of a parallel-tuned circuit as a function of circuit "Q."

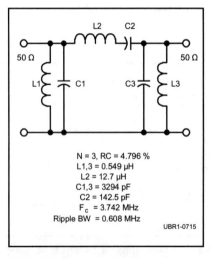

N = 3, RC = 4.796 %
L1,3 = 0.549 μH
L2 = 12.7 μH
C1,3 = 3294 pF
C2 = 142.5 pF
F_c = 3.742 MHz
Ripple BW = 0.608 MHz

Fig 7-15—Representative three-stage band-pass filter.

As with low and high-pass filters, the method by which we can get the skirts and the passband width we want is to use multiple tuned circuits, generally with slightly different resonant frequencies. This technique, called *stagger tuning*, results in a relatively smooth passband, wider than with a single sharp section, and steep skirts with each filter section helping to reduce the level of signals away from the desired passband. A filter consisting of two parallel tuned circuits coupled by a series tuned circuit is shown in **Fig 7-15**.

To realize a passband sharper than a single tuned circuit, we may tune all to the same frequency. If we wish to achieve a passband wider than any one of the sections, we might set the series-tuned circuit near the middle of the desired passband and the two parallel circuits near the edges. Depending on how far apart the edge frequencies are, we can achieve a fairly flat passband response, as shown in **Fig 7-16**.

Piezoelectric Crystal Filters

In our discussion of oscillators in Chapter 4, we discussed the piezo-electric crystal, a slab of quartz (see Fig 4-6) that is precisely shaped to respond to signals at a particular frequency. Not surprisingly, the same kind of response can be used for filtering. A quartz crystal of this type has a much higher Q than circuits of inductors and capacitors, and thus a crystal-based filter can be quite effective if applied to the right kind

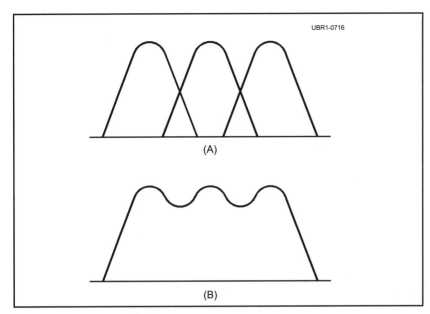

Fig 7-16—At A, response for each tuned circuit in Fig 7-15, and at B, the combined response.

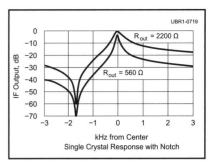

Fig 7-19— Response of a single-crystal filter with holder capacitance purposely not balanced. Shown for two values of R_{out}.

of signals. A single crystal can be used as a filter in the circuit of **Fig 7-17**. The crystal acts like a series-resonant circuit, passing signals at its resonant frequency. The holder that the quartz is mounted in acts like an additional capacitance across the crystal. Note the capacitor C_C in the circuit. If this is set to cancel the capacitance of the holder, the response is that of a high-Q series-resonant circuit, as shown in **Fig 7-18**. If the capacitance is not cancelled out, it results in both a series and a parallel resonance at slightly different frequencies. The response in this case is quite interesting, as **Fig 7-19** shows. Note that this response both selects a desired signal and notches a nearby undesired one, a useful arrangement in some cases, as we'll discuss later.

For many kinds of signals, the sharpness of a single crystal filter is too sharp to pass the required signal bandwidth. By using multiple crystals in a *lattice* configuration, a response with a wider passband can be achieved, while still maintaining the steep skirts inherent in the crystal response. The simplest dual-crystal circuit is shown in **Fig 7-20**. Note that with two crystals in the same size and type holder, the capacitance of each balances out the other, so

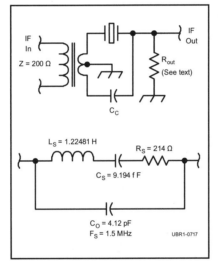

Fig 7-17—Circuit of a single-crystal filter for CW.

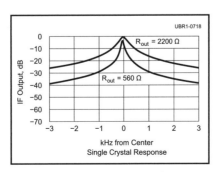

Fig 7-18—Response of a single-crystal filter with holder capacitance balanced for CW use. Shown for two values of R_{out}.

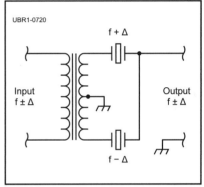

Fig 7-20— Circuit of a two-crystal filter.

only series resonances are involved. The response of two crystals 1.5-kHz apart, each with a frequency of around 1.5 MHz is shown in **Fig 7-21**. As in the case of LC filters, often many additional stages are used to obtain steeper skirts, with configurations of up to eight crystals commonly used.

Mechanical Filters

In the early 1950s, engineers at Collins Radio Company (now a part of Rockwell-Collins) developed a radical kind of band-pass filter called a *mechanical filter*. The Collins mechanical filter used a *transducer*, a device to convert electrical signals into mechanical movement, very much like a tiny loudspeaker, to drive a row of metal disks. The disks were held apart, but coupled together by small rods along their edges. The output at the far end of the row was

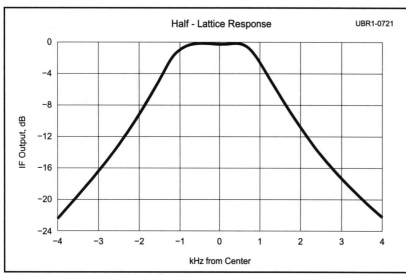

Fig 7-21—Response of a two-crystal filter for SSB use.

passband and steeper skirts than filters of other types of the day. Typical center frequencies ranged from 200 to 500 kHz and bandwidths ranged from 500 Hz to 16 kHz, covering most of the communication signal bandwidths of the day. In the next chapter we will discuss how we will move higher frequency signals to allow them to be processed by such filters.

In recent years, crystal-lattice filters have become available that have similar response to the mechanical filters. But in their day, mechanical filters were the best available. They are still produced today and used in many types of equipment. New mechanical filters are smaller—closer to the diameter of a cigarette and half as long. Mechanically resonant filters of other materials (such as ceramics) are sometimes encountered as well.

converted back to an electrical signal by a second transducer. The whole device was narrower than a cigar and half as long.

The disks were designed to be mechanically resonant and to vibrate over a narrow frequency range. The result was a filter with a flatter

Review Questions

1. Design a single element, passive RC high-pass filter with a load impedance of 600 Ω and a cut-off frequency of 3000 Hz. Evaluate the transfer function in voltage and dB at 100, 3000, 6000 and 100,000 Hz.

2. If we wish to filter a 10-kHz AM broadcast signal at 500 kHz, 1.5 MHz and a SW broadcast signal at 10 MHz, what percentage of bandwidth are we talking about at each carrier frequency?

3. If we were to use a 7-kHz wide filter, instead of a 10-kHz filter to process an AM broadcast signal, what would be the result?

Chapter 8

Active and Digital Filtering

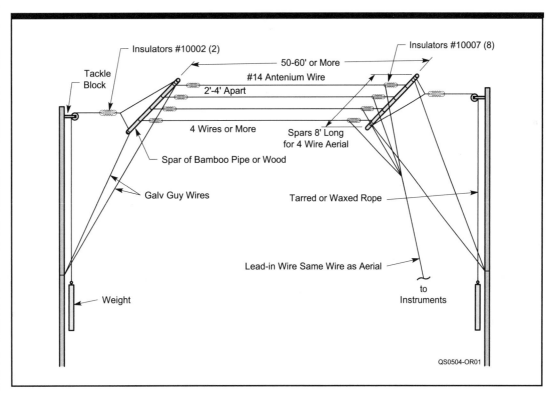

A well-made spark station back in the early 1900s had an antenna system like this!

Contents

A special case of filters is the *active filter*. The filters we've talked about so far have been constructed of *passive* elements, such as inductors, capacitors and resistors. An active element is one that provides gain, operates with external applied power and generally has three or more terminals, such as an amplifier module or integrated circuit. An active filter is based on an *operational amplifier*, an amplifier building block originally developed for analog computers, something we don't see very much anymore. An ideal operational amplifier uses feedback to determine its characteristics and is assumed to be an infinite gain, infinite input impedance, zero output impedance device when feedback isn't applied. Of course, no amplifier has these ideal characteristics, but a practical *op-amp* is close enough that it makes the analysis work. Op-amps can be used for many functions and are often seen as basic dc, audio or RF amplifiers. A simple "all-pass" operational amplifier with feedback is shown in **Fig 8-1**. Note the expression for the inverted amplifier gain, just equal to the value of the feedback resistor divided by the input resistor. It is independent of the actual gain of the op-amp as long as the op-amp gain without feedback is much larger than the design gain when feedback is applied.

To make this amplifier into an active low-pass filter, we merely connect a capacitor in parallel with the feedback resistor, as shown in **Fig 8-2**. At very high frequencies, the impedance of the feedback

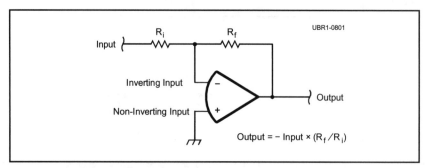

Fig 8-1—Operational amplifier with feedback (power connections not shown).

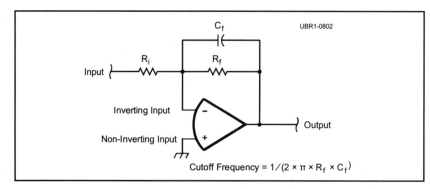

Fig 8-2—Operational amplifier with feedback becomes an active low-pass filter with the addition of a capacitor.

network approaches zero and there is no output, while at very low frequencies the impedance of the feedback network is just that of the resistor and we get full gain. At the frequency at which the capacitive reactance equals the resistance of the feedback resistor, the total reactance is 0.707 of the resistor and thus the output is at the –3-dB point, also known as the cutoff frequency.

The active filter has a number of benefits. First, unlike a passive filter that always has loss of some amount,

the active filter can simultaneously function as an amplifier *and* a filter. Next, it allows a large amount of flexibility in selection of values— You can adjust the resistance values to allow use of a convenient capacitor value. Finally, very effective filters can be constructed without inductors, usually the bulkiest and often most expensive components. As you would expect, low-pass and band-pass configurations are possible in addition to the high-pass filter we just discussed.

Digital Filtering

It should come as a surprise to no one, but the digital revolution is upon us! Many tasks successfully completed in an analog fashion for many years are now done more effectively and efficiently with digital processing. While some resist this trend, there is no question that digital processing can do a good job, especially if we understand the design guidelines and limitations—Filtering is no exception.

While general-purpose digital computers can be made to filter signals, most successful digital filtering is carried out using *digital signal processors* or DSPs. DSPs perform the same general tasks as general-purpose processors; however, they are especially designed to process signals in real time at high speed. While many functions can be performed, we will concentrate here on the filtering process.

To process signals digitally, the first task required is to convert our analog signal into a digital one. This is accomplished with a *digital to analog converter* or DAC. The DAC establishes a key parameter of the process—the *dynamic range* of the signals the processing can handle. This is the range of amplitude of the signals the processing must handle. In an analog filter, there is a dynamic range too, but it's very wide. At some high level, we could crack a crystal or melt an inductor, but our intuition usually keeps us on safe ground. With a DAC, the number of bits that the DAC converts the data to sets the range. We can adjust the process so that the smallest signal we expect to deal with can be coded into a few bits, and then the total number of bits determines the maximum signal we can handle. The dynamic range is the difference between the two.

For example, if we have a 16-bit DAC, the maximum value is just 2^{16} or 65,536 above the size of a single bit. Expressed in dB, this is 96 dB.

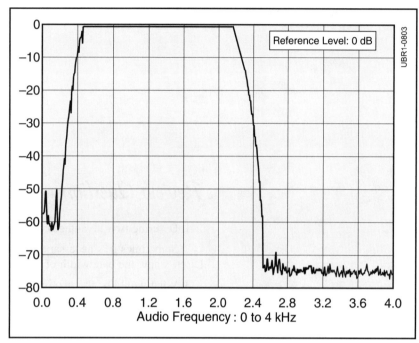

Fig 8-3—Response of a commercial audio digital filter. (SGC ADSP² in voice mode).

This is comparable to other receiver dynamic ranges, but by no means up to the state of the analog art. DSPs are available, however, with 24 or even 32-bit sampling and processing, so dynamic range needn't be a problem. You just need to be aware that it can be an issue.

The DAC sampling and subsequent processing must occur rapidly enough to represent signals that are changing. You can't process a signal if you don't capture the way it's changing.

In 1924, long before people were talking about computer data, Henry Nyquist, an AT&T engineer, espoused his famous theorem, now the basis for digital telecommunications, stating that you needed a minimum of two samples at the highest frequency you wished to process. This is very much like a motion picture camera that takes a snapshot 30 times a second. Our eyes and brain turn things back into a continuous view.

Let's say we want to process the audio signal coming out of our crystal set. (Now adding a complicated Digital Signal Processor on the output of a simple crystal set does give this author pause!) If we are listening to broadcast stations with frequency components up to 5 kHz, we would need to use a DAC that sampled at least twice that, or 10,000 times a second. That is something that we could buy off-the-shelf at a reasonable price. On the other hand, if we wanted to replace a 500-kHz mechanical filter with a DSP based digital filter, we would need one that sampled at 1 MHz minimum. That is not yet readily available in 2005, but it's not far over the horizon.

You will perhaps be relieved to know that in this book, we don't intend to dig deeply into either the theory or the software involved in digital filtering. We can briefly mention that a mathematical process known as differential equations can exactly define the detailed time response of LC filtering, in which the rate of change of voltage of capacitors and inductors over time can be used to predict their behavior. The digital processing equivalent is known as *difference equations*. The predicted behavior during one sample time is compared with the value of the signal at the next sample and a similar response can be predicted.

With DSPs, it is possible to perform this many more times than is possible with an analog filter, simulating many cascaded stages, with the result that the filter response becomes very close to the idea shapes described earlier in the chapter. See **Fig 8-3** for an example. Expect to see digital filters in more and more applications in the future.

Review Questions

1. Describe two advantages of active filters compared to passive filters.

2. Sometimes it is necessary to have multiple bandwidths available. Discuss how the bandwidth of active and digital filters could be changed.

3. What would be the maximum dynamic range for voltage samples from 8, 16 and 24-bit DACs?

Chapter 9

Heterodyne Receivers—Putting the Pieces Together

Sometimes a group of hams will get together to operate from a rare and exotic location. This is called a "DXpedition," this one to Rodriques Island in the Indian Ocean, deriving from the term "DX" used by hams to describe a distant contact.

Contents

In the early days of radio, all transmission was done with the dots and dashes of Morse code. The earliest *spark gap* and later *rotary gap* transmitters were not just on-and-off keying, as described in Chapter 1. They also included the modulating frequency of the buzzing interrupter of the spark gap or the more musical note of a higher-pitched rotary gap. These allowed the receiving operator to hear the key-down *on* periods so an operator listening on a simple crystal set could understand the code characters. Contrast this with our earlier example in Chapter 2 of on-off data transmission. There the received pulse train would not be audible to someone listening with a crystal set.

When radio moved from the spark era to vacuum-tube transmitters sending continuous rather than the interrupted waves of the spark transmitters, operators had a problem hearing the tone of such clean signals! One early solution was to put enough feedback around an RF amplifier stage such that it would just oscillate at a frequency slightly above or below the desired signal. The stage would simultaneously amplify the desired signal and generate a local signal to mix with the incoming one. The signals would drive the amplifier into a non-linear region, making it act like a mixer, and the difference frequency would go to an audio amplifier, resulting in an audible tone replica of the received signal that was only present when the key was down.

The resulting receiver was called a *regenerative* receiver. It was refined into a reasonable performer and was in popular use for the reception of continuous-wave (CW) radiotelegraph signals from the 1920s through the 1930s. AM voice could be received by setting the gain just below the point at which the receiver started oscillating. Even though usable, the regenerative receiver suffered from a number of problems, including all the limitations of the TRF, since it was just a TRF (see Chapter 3) with a different form of detector. In addition the regenerative detector was very touchy to keep on the correct frequency since the gain and tuning tended to interact. It also generated a signal that could be radiated by the antenna, resulting in interference to other users of adjacent frequencies unless steps were taken to isolate the detector from the antenna.[1]

A better approach is the *heterodyne* receiver. This performs the same functions as the regenerative receiver above but separates the pieces, eliminating some of the problems. The heterodyne receiver makes use of separate oscillator, mixer (detector) and amplifier, as shown in the block diagram of **Fig 9-1**. By not trying to use the amplifier both as amplifier and local oscillator, as in the regenerative receiver, the heterodyne receiver's circuits can be optimized for each function. One simple heterodyne configuration that is sometimes encountered today is called a *direct-conversion* receiver.

Finally—The Superheterodyne!

The *superheterodyne*, or *superhet* for short, uses the principles of the heterodyne above—twice. In so doing, it neatly sidesteps the problems of both the TRF and the regenerative receiver. In the superhet, a local oscillator and mixer are used to translate the received signal to an *intermediate frequency* or IF, for additional amplification and processing. Then a second mixer is used to detect the IF signal, translating it to audio. The configuration is shown in **Fig 9-2**.

In a typical configuration, the local oscillator (LO) and RF amplifier stages are adjusted so that as the local oscillator is changed in frequency, the RF amplifier is also tuned to the appropriate frequency to receive the desired station. An

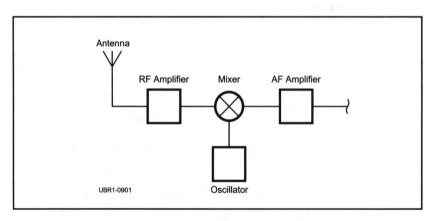

Fig 9-1—The heterodyne, or direct-conversion, receiver.

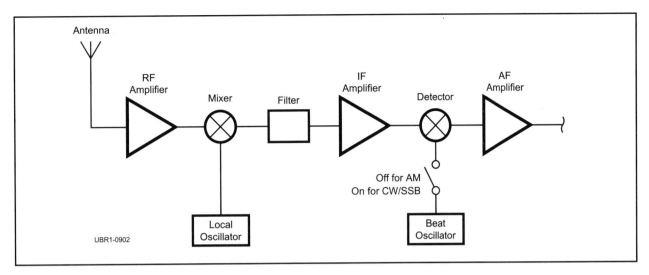

Fig 9-2—A block diagram of a superheterodyne receiver.

example may help. Let's pick a common IF frequency used in an AM broadcast radio, 455 kHz. Now if we want to listen to a 600-kHz broadcast station, the RF stage should be set to amplify the 600-kHz signal and the LO should be set to 600 + 455 kHz = 1055 kHz.[2] The 600-kHz signal, along with any audio information it contains, is translated to the IF frequency and is amplified. It is then detected, just as if it were a TRF receiver at 455 kHz.

Note that to detect standard AM signals, the second oscillator, usually called a *beat frequency oscillator* or *BFO*, is turned off since the AM station provides its own carrier signal over the air. Receivers designed only for standard AM reception generally don't have a BFO at all.

It's not clear yet that we've gained anything by doing this; so let's look at another example. If we decide to change from listening to our station at 600 kHz, we can tune the single dial of our superhet to 1560 kHz. With the appropriate ganged and tracked tuning capacitor, the LO is set to 1560 + 455 = 2010 kHz and now that station is translated to our 455-kHz IF.

Note that the bulk of our amplification can take place at the 455-kHz IF frequency, so not as many stages must be tuned as we change to a new frequency. Note also that with the superheterodyne configuration the

selectivity (the ability to separate stations) occurs in the intermediate-frequency stages and is thus the same no matter what frequencies we choose to listen to. The superhet design has thus eliminated the major limitations of the TRF, at the cost of two additional building blocks.

It should be no surprise then that the superhet, in various flavors that we will discuss, has become the primary receiver architecture in use today. The superhet concept was introduced by Major Edwin Armstrong,[3] a US Army Artillery officer, just as WW I was coming to a close. The superhet quickly gained popularity and following the typical patent battles of the times became the standard of a generation of vacuum-tube broadcast receivers. These were found in virtually every US home from the late 1920s through the 1960s, when they were slowly replaced by transistor sets, but still of superhet design.

The superhet has withstood the test of time and is the basic architecture in radios covering all sorts of frequencies and applications—from standard broadcast sets to television sets, microwave radios and other communications systems, radar systems and even cell phones. Now we will discuss some different operating modes that require different bandwidths and how to implement them in a superhet.

Designed for Application—Signal Bandwidth

In the last section we discussed the evolution of the superhet receiver, starting from the era of the crystal set. In this section we will discuss the requirements of different applications and the resulting details of superhet designs.

Earlier, I mentioned that one advantage of a superhet is that the bandwidth is established by the IF portion. It is thus independent of the RF frequency to which the receiver is tuned. It should not be surprising that the detailed design of a superhet receiver is dependent on the nature of signal being received.

Signal Bandwidth—an Example—Amplitude Modulation (AM)

In Chapter 2 we developed the concept that if we multiplied (in other words, modulated) a 600-kHz carrier with a 600-Hz tone, we would generate additional new frequencies at 599.4 and 600.6 kHz, as shown in **Fig 9-3A**. If instead we were to modulate the 600-kHz carrier signal with a band of frequencies corresponding to toll-quality human speech of 300 to 3300 Hz, we would have a pair of bands of information carrying waveforms extending from 596.7 to 603.3 kHz, as shown in **Fig 9-3B**. These bands are called *sidebands,* and some form of these is

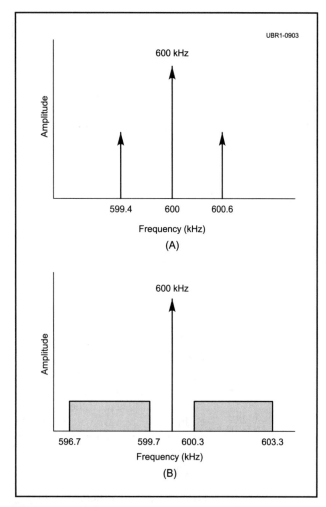

Fig 9-3—At A, frequency spectrum of a 600-kHz carrier modulated by a 600-Hz tone. At B, frequency spectrum of a 600-kHz carrier modulated by speech (300 to 3300-Hz).

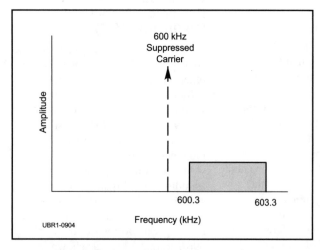

Fig 9-4—Single (upper) sideband of the AM speech signal sent without a carrier.

present in any signal carrying information.

Note that the total bandwidth of this AM voice signal is twice the highest frequency transmitted, or 6600 Hz. If we choose to transmit speech and limited music, we might allow modulating frequencies up to 5000 Hz, resulting in a bandwidth of 10,000 Hz, or 10 kHz. This is the standard channel bandwidth that commercial AM broadcasters are permitted to occupy in the US. We would refer to this as a *narrow-bandwidth* mode.

What does this say about the bandwidth needed for our receiver? If we want to receive the full information content transmitted by a US AM station, then we need to set the bandwidth to 10 kHz. What if we have a narrower bandwidth? Well, you will lose the higher frequency components of the transmitted signal, perhaps ending up with a radio suitable for voice but not very good at reproducing music.

On the other hand, what is the impact of having too wide a bandwidth in our receiver? In that case, you will be able to receive the full transmitted spectrum but you will also receive some of the adjacent channel's information. This will sound like interference and reduce the quality of what you are receiving. If there are no adjacent channel stations, you will get any additional noise appearing in the additional bandwidth, but no additional information. The general rule is that the received bandwidth should be matched to the bandwidth of the signal you are trying to receive.

Another Choice—Single Sideband Suppressed Carrier Modulation (SSB)

The standard commercial AM format is very convenient for receiver design, since the carrier needed to demodulate the received sidebands is sent along with the sidebands. Applying the total received signal to a detector or multiplier, such as in our crystal set, allows the audio to be recovered without having to worry about any of the finer points we will discuss later. This is very cost effective in a broadcast environment when there are many receivers and only a few transmitters.

In looking at Fig 9-3, you might have noticed that both sidebands carry the same information, and are thus redundant. In addition, the carrier itself conveys no information. It is thus possible to transmit a *single sideband* and *no carrier*, as shown in **Fig 9-4**, relying on the BFO (beat frequency oscillator) in the receiver to provide a signal to multiply the sideband with to provide demodulated audio output. The implications in the receiver are that the bandwidth can be slightly less than half that required for double sideband AM (DSB). There must be an additional mechanism to carefully replace the missing carrier within the receiver. This is the function of the BFO, which must be at just exactly the right frequency. If the frequency is improperly set, even by a few Hz, James Earl Jones can come out sounding like Donald Duck!

This makes a requirement for a much more stable receiver design, with a much finer tuning system. Such a receiver thus is more expensive. An alternate is to transmit a reduced carrier and have the receiver lock its tuning to the weak carrier, usually called a *pilot* carrier. Note that the pilot carrier need not be of sufficient amplitude to demodulate the signal, just enough to allow a BFO to lock to it. These alternatives are effective, but tend to make SSB receivers expensive and most appropriate for the case in which a small number of receivers are listening to a single transmitter, as is the case of most two-way communication systems.

Radio Telegraph

I have described radiotelegraphy in earlier chapters as being transmitted by "on-off keying of a carrier." You might think that since a carrier takes up just a single frequency, the receive bandwidth needed should be almost zero. This is only true if the carrier is never turned on and off. In the case of telegraphy using Morse code, it can be turned on and off quite rapidly. The rise and fall of the carrier results in sidebands extending out from the carrier for some distance, and they must be received in order to reconstruct the signal in the receiver.

A rule of thumb is to consider the rise and fall time as about 10% of the pulse width and the bandwidth as the reciprocal of the quickest of rise or fall time. This results in a bandwidth requirement of about 50 to 200 Hz for the usual radiotelegraph transmission rates. Another way to visualize this is with the bandwidth being set by a *high-Q* tuned circuit. Such a circuit will continue to "ring" after the input pulse is gone. Thus too narrow a bandwidth will actually "fill in" between the code elements and act like a no-bandwidth, full-period carrier.

Data Communication

Data Communication—Radio Teletype

One of the first successful machine telegraph systems was the *Teletype*. This innovative system, when new in the late 1800s, allowed data (or then more properly called *record*) communication to be carried out by anyone who could operate a typewriter. This system made use of a five-unit code-named after its inventor, Emile Baudot, to send letters and numbers over a telegraph circuit. With five elements, the maximum number of characters that can be sent is $2^5 = 32$. This causes a problem for our alphabet with 26 letters, ten numerals and a number of punctuation and special characters. Baudot decided that instead of having separate characters for numbers and punctuation marks, he would assign a special character to switch between *letters* and *figures*. This results in each key or code element being able to be used for *two* characters, although extra characters must be sent to switch between the two functions. The Baudot code always includes a start and a stop pulse to indicate the beginning and end of each character. This was necessary since a character could start whenever an operator struck a key on the keyboard. **Fig 9-5** shows the letter "A" in Baudot code, with start and stop pulses. Note that the stop "pulse" is really an undefined interval until the operator hits another key.

More recent codes, such as the common ASCII code (American Standard Code for Information Interchange), use seven elements to provide $2^7 = 128$ possible characters. This is enough for lower- and upper-case letters (26 each), numerical digits (10), punctuation codes (30 on a computer keyboard) and special characters associated with various languages, as well as non-printing control characters. ASCII characters may also have an error-indicating parity bit appended, making them into a computer-convenient eight-bit code. They also may have start and stop characters, as with Baudot code, if used in keyboard applications. The ASCII code of the letter "a" is shown in **Fig 9-6**.

While the Baudot code can be sent using on-off keying, the ability to detect characters in the presence of noise is improved by having a well-defined state for both a Baudot *mark* (named after the early Morse tape readers that placed a pen mark on a paper tape when the key was down) and *space*, the state where the key is up. In wire-line transmission, the line current is reversed between a mark and space.

This well-defined, two-state encoding was carried over to radio transmission of Baudot radio teletype (RTTY). It is thus customary to send two distinct tones to carry RTTY: one for mark and one for space. This is referred to as *frequency shift keying* (FSK) and is the simplest form of *frequency modulation* or FM. The transmitter sends one frequency for a mark; the other for a space, shifting back and forth as the typist's keystrokes generate each five-pulse burst. Each tone can be considered as a separate Morse-like stream as discussed above, except the speeds are higher, ranging from 65 to 133 words per minute, depending on the standard used.

Baudot transmission is being replaced by more modern coding, such as ASCII, which can be readily derived from computers or computer terminals. Amateur Radio operators and some other services[4] still continue to use Baudot and typically use a 170 to 850-Hz separation between the mark and space frequencies, depending on the data rate and local convention. The minimum bandwidth to recover the data is around twice the spacing between the tones. Note that the tones could be generated by directly shifting the carrier frequency, or by using a pair of 170-Hz spaced audio tones applied to the audio input of an SSB transmitter. If the frequencies are appropriately adjusted, both will appear the same to a distant receiver.

Even if the standard audio tones of 2125 Hz (mark) and 2295 Hz (space) are used, the required bandwidth is still around 340 Hz (for 170-Hz shift), even though the tones are quite a bit removed from the (suppressed) carrier frequency. Note also that the tones fit within the bandwidth of a voice channel and thus the facilities of a voice transmit-

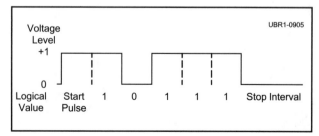

Fig 9-5—The Baudot code for the letter "a".

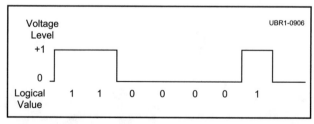

Fig 9-6—Voltage pulses corresponding to ASCII code of letter "a".

ter and receiver can be employed without any additional processing needed outside the radio equipment. Alternately, the receiver can employ detectors for each frequency and provide an output directly to a Teletype machine or terminal.

Data Communication—ASCII Transmission

As I mentioned above, modern computer codes can be accommodated in a similar manner to RTTY, with a tone used for each state of the data pulses. Using a pair of tones as in RTTY results in a maximum data rate of about 1200 bps over a voice channel. However, the error performance is quite good in the presence of noise. If the channel performance is solid (there is a high *signal-to-noise ratio*, described later), standard telephone voice-modem techniques can be used, making use of multiple carriers, multiple levels or multiple phase shifts (more generally a combination of these techniques) to achieve a data throughput of close to 30,000 bps over a voice-bandwidth channel.

Sending Voice as Data — Pulse Code Modulation

Transmission and processing systems designed for data applications can also be used for voice and other analog signals if they are encoded into data-type words. The first successful, and still most common, format was called *pulse code modulation* (PCM). The conversion of analog signals into digital format occurs in two steps.

- **Sampling**: A snapshot of the level of the analog signal is taken at regular intervals. In 1924, long before people were talking about computer data, The Nyquist

theorm, described in Chapter 8 requires that if at least two samples are taken at the highest frequency of a signal, the signal could be accurately reproduced from the samples. To sample voice signals up to 3300 Hz, you need to sample at twice that rate or at least 6600 times a second. Even with the best filters of that day, some higher frequency signals passed through, so 8000 samples per second was established as the standard voice-sampling rate.

- **Encoding**: To transmit a digital signal, each sample must be encoded into a digital word representing the value of that sample. The accuracy of the encoding is dependent on the number of bits we allocate to each sample. For example, if we just have one bit, about the best we can do is specify whether a sample is positive or negative. If we have two bits, we can encode as four values: 00, 01, 10 and 11. With three bits there are eight values. In general, we can have 2^n values for n bits.

Let's look at the data standard of eight bits (a Byte to computer folks). With eight bits, you can specify a sample as being one of 256 values. If your voice signal can be between ±1 V, you can specify the value to the nearest $2/256 = 0.01$ V. The difference between the real value and the actual one can be considered a kind of noise. Compared to the maximum value this amount of noise represents a *signal-to-noise ratio* (SNR) of 40 dB.

If we generate and send one of these eight-bit coded words 8000 times a second, we end up with a PCM throughput of 64,000 or 64 kbps. If we just send these bits over a digital radio, each bit time is

$1/64000 = 15.6$ msec. Using the 10% rule, this requires a bandwidth of about 640 kHz, well beyond our usual narrow-bandwidth criterion. However, with appropriate processing, compression and modem techniques—making use of multiple tones per signal—a digitized voice channel can be made to just fit in an analog voice channel slot.

So why go to all this trouble, when a voice channel fits very well in an analog voice channel? One reason is SNR. We noted above that the effective SNR of digital voice is based on the encoding. So long as we can detect whether the data bits corresponding to the voice samples are a one or zero, we will have a SNR of the 40 or so dB described above—even if the actual transmission SNR drops to perhaps 10 dB. Another reason is that if we require privacy, we can use powerful digital-encryption techniques to encode the data.

The primary use of PCM, however, is as a part of a multi-channel voice multiplex system. This will be described in the next chapter.

Notes

[1]It took a while for the US Navy to determine that these signals were providing a direction-finding beacon for enemy U-boats!

[2]The local oscillator (LO) could also be set to 600 − 455 kHz = 145 kHz. However, this choice would make it more difficult to cover the entire band.

[3]He is the same Armstrong who invented Frequency Modulation (FM) some years later and who held many radio patents between WW II and WW I.

[4]The Baudot code is still the standard employed in wire-line teletype (TTY) keyboard terminals used by hearing-impaired people.

1. What frequencies can a superheterodyne receiver hear if the intermediate frequency is 1 MHz and the local oscillator is set to 4 MHz? How can you make it receive only one of the signals?

2. It is desired to transmit music with frequency components from 50 to 10,000 Hz. Compute the bandwidth required to transmit the resulting signal using AM (DSB) or SSB.

3. Discuss benefits of each of the above modulation schemes.

4. List all the eight bit patterns that correspond to the allowed encoding of a three-bit code word.

Chapter 10

The Modern Superheterodyne

The "Squad 14" high-tech communications bus used by the Kodiak Amateur Radio Emergency Services (KARES) volunteer fire and rescue group in Alaska.

Contents

The discussion in the preceding section related to transmissions typically employed in the MF, HF and the lower VHF ranges. In that portion of the radio spectrum, bandwidth is generally limited to modulation methods occupying bandwidths comparable to a normal voice channel. National and international regulatory bodies impose such restrictions, taking into account the limited bandwidth available and the international nature of radio signals, particularly in the HF region.

Above the lower VHF regions, wider bandwidths are often used. Signals tend to stay confined in a relatively small geographic area, so channels may be reused in other areas. This, provides more effective use of the radio spectrum. Another factor to consider is that the total bandwidth in VHF (30 to 300 MHz) is more than nine times the bandwidth of the entire VLF, LF, MF and HF ranges combined (30 kHz to 30 MHz).

While narrow-band transmissions, such as the ones found in the lower-frequency regions, are frequently used at VHF and above, it is also common to encounter wider bandwidths. Some examples are those used by wideband FM, television, pulsed radar and multiplexed multi-channel transmissions. We will briefly discuss each—and their implications on receiver bandwidth—below.

Frequency Modulation (FM) Transmission

The simplest form of Frequency Modulation (FM) was introduced under the topic of *Data Transmission—Radio Teletype* in Chapter 9. FM can also be used to carry voice and other kinds of information. Unlike AM, in which the amplitude is varied to convey the information about the magnitude of the modulating signal, with FM the frequency is

changed. The loudest voice peaks in the amplitude of the modulating signal result in the maximum change in the carrier frequency.

The frequency is said to *deviate* with the modulating signal. The maximum amount that the carrier signal is allowed to shift is called the *frequency deviation*. You might expect that the required bandwidth would be the difference between the carrier minus the deviation, and the carrier plus the deviation, which is twice the deviation. For real modulating signals, the required bandwidth is somewhat greater, and is approximately equal to 2 (M + D) where D is the deviation as described above and M is the highest modulating frequency. If we have 3.3 kHz voice and a deviation of 3.3 kHz, the required bandwidth would be 13.2 kHz, or just twice that of standard AM for the same signal.

The ratio of deviation-to-maximum-modulating-frequency is called the *modulation index*. It can be shown that the modulating index must be at least one, but it can be higher. When the modulation index is small, typically less than two, it is

referred to as *narrowband FM*, while a modulation index of five or higher is referred to as *wideband FM*. Wideband FM can be shown to have an improved signal-to-noise ratio over narrowband FM (or AM), all other things being equal, and is used for high-quality music transmission. In the FM broadcast band (88 to 108 MHz in the US) a 150-kHz bandwidth channel is assigned to each station. If mono (rather than stereo) transmission is employed, a frequency response to 20 kHz can be supported while using a modulation index of five.

The simplest detector for FM is just to use the response of an AM receiver with sufficient bandwidth—as noted, even narrowband FM is wider than AM—and the response of a tuned circuit as shown in **Fig 10-1**. As the signal deviates back and forth across the slope of the tuned circuit, the output voltage goes up and down, reproducing the modulating signal.

By combining two resonant circuits and two amplitude detectors, it is possible to have an FM detector in which the center frequency is in the center of the passband. Such an

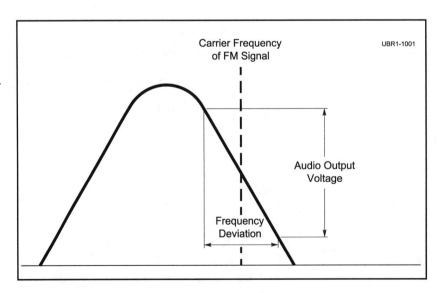

Fig 10-1—Using a parallel-resonant circuit as an FM detector.

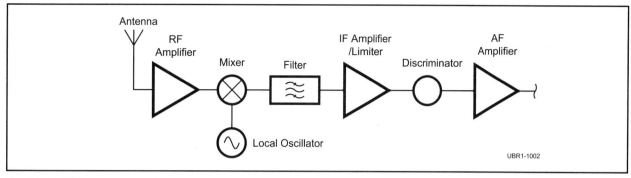

Fig 10-2—Block diagram of an FM radio receiver.

FM detector is called a *discriminator* and is probably the most common type encountered in FM systems.

Because the FM receiver responds to frequency rather than amplitude variations, it is possible to eliminate amplitude noise added to the signal using a *limiter*. The limiter is just an amplifier that is biased so that the positive and negative peaks of the signal waveforms are clipped, and therefore eliminated at the output. A block diagram of a basic FM receiver is shown in **Fig 10-2**. As you may observe, it is quite similar to an AM receiver. Some receivers are designed to receive both FM and AM signals by switching from an AM detector to limiter(s) and a discriminator. Note that for best performance in each mode, different IF bandwidth filters are needed as well.

Multi-Channel Multiplex Transmission

If we consider a SSB transmitter, we can imagine that nothing prevents us from simultaneously transmitting multiple SSB transmissions using the same transmitter at the same time. Some early implementations used an AM channel to transmit two SSB channels, one using the upper and the other the lower sideband. This transmission method is called *Independent Sideband* or ISB, and allows two different voice transmissions to be sent, one on either side of a suppressed carrier.

While ISB is fine for HF transmission of two voice signals in the bandwidth of a single AM voice channel, the concept has been

significantly extended for VHF and microwave use. There is nothing that prevents multiple SSB voice channels from being stacked up above (or below) a suppressed-carrier frequency in order to carry many voice circuits on a single wideband radio channel. This was the method used by telephone companies to provide long-distance voice service by microwave radio until the fiber-optic revolution starting in the mid 1980s.

If you allocate 4 kHz of bandwidth to a voice channel (easy if you are using audio frequencies from 300 to 3300 Hz), you can group twelve channels into a voice channel (VC) *group* with a bandwidth of 48 kHz. By convention, this group will extend from 60 to 108 kHz. A block diagram of the transmit side of such a *frequency division multiplexer* (FDM) is shown in **Fig 10-3**. This could be applied directly to the

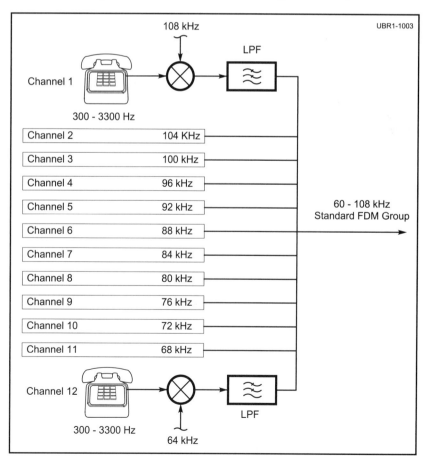

Fig 10-3—Frequency-division multiplex 12-channel group.

Table 10-1

Worldwide digital-voice hierarchies used in radio transmission.

NORTH AMERICA			JAPAN			EUROPE		
Name	*Rate*	*VC*	*Name*	*Rate*	*VC*	*Name*	*Rate*	*VC*
T0	64 Kbps	1	J0	64 Kbps	1	E0	64 Kbps	1
T1	1.544 Mbps	24	J1	1.544Mbps	24	E1	2.048 Mbps	30
T1C	3.152	48						
T2	6.312	96	J2	6.312	96	E2	8.448	120
T3	44.736	672	J3	32.064	480	E3	34.368	480
T4	274.176	4032	E4	139.264	1920			

modulator of an SSB transmitter and sent to the next telephone office. In the microwave region, the phone company radio channel assignments are significantly wider than 48 kHz, so the groups have been multiplexed up with four other groups to become a *supergroup* of 240-kHz bandwidth carrying 60 channels, or a *mastergroup* of five supergroups at 1200 kHz (1.2 MHz) carrying 300 voice channels, topping out at a *supermastergroup* of three supergroups requiring 3.6 MHz of bandwidth and carrying 900 voice channels.

Such microwave-transmission systems formed the backbone of the US national voice network until the end of the analog telecommunications era, just before 1990. While appearing massive to those thinking in terms of a single voice channel on a radio circuit, each of the 900 channels was just an SSB channel of the type we described earlier, one stacked on top of another. While this arrangement worked well, fiber worked better for this application and telephone microwave systems in the US have been largely replaced by digitally encoded fiber-based systems. The analog technology is still viable and is used in many areas that are not hospitable to buried fiber systems or those that were in place and don't need the added capacity available from fiber.

The PCM encoding described in the previous chapter has now replaced most analog multiplexing. In typical use, multiple 64-kbps channels are combined, by inserting one eight-bit sample from a channel after that from another to make up a *frame*. In North America and Japan 24 such channels (1.536 Mbps total) are combined with a synchronizing bit to make up a data frame (1.544 Mbps) called a T1 in NA and a J1 in Japan. In Europe, 30 PCM voice channels are combined with a signaling channel and a synchronizing channel to make a 32-channel frame, called an E1 with a resulting throughput of 2.048 Mbps.

Each of the standards provides for higher-level multiplexing through bit-by-bit interleaving of the PCM data streams for higher throughput systems. The levels used in radio systems are summarized in **Table 10-1**. Higher-throughput systems are defined and used for fiber-based systems, with modern fiber transmission systems carrying Gbps data on each light carrier and wavelength-division multiplexing used to combine multiple light carriers on a single strand. The fact that fiber is relatively inexpensive (compared to the cost of digging a trench to put it in) has made long-haul telecommunications costs drop dramatically.

Baseband—The Information Signal

In many system designs, such as the multi-channel voice systems described above, the equipment used to combine the individual channels (usually called a multiplexer) is separate from the radio equipment. In this way, the same multiplexer can be used with different radio equipment, or even for direct transmission via terrestrial coaxial or undersea cable. The radio and multiplex equipment share a common set of interface standards so that different manufacturers' equipment can be selected to optimize the combined system for a given application.

The signal that crosses the boundary between the multiplexer and the radio is called a *baseband* signal. Radios are specified to accept baseband inputs composed of different kinds of signals and don't really care if they come from voice, data or image sources, so long as they occupy a bandwidth equal to or narrower than the transmit bandwidth. People who design and manufacture such equipment tend to focus on different parts of the spectrum and have different but appropriate skills for either job.

Television (TV) Transmission

Each North American standard broadcast television signal is assigned a bandwidth of 6 MHz. The signal is referred to as a *composite-video* signal since it contains a number of separate signals within the same channel. The channel spectrum is shown graphically in **Fig 10-4**. Note that the bulk of the channel, more than 5 MHz, is comprised of video information. This is transmitted via an amplitude-modulated carrier located 1.25 MHz above the bottom of the channel. This is sent as double sideband AM, however, the sidebands are not equal—the higher-frequency components are attenuated in the lower sideband. The sound is carried via a separate FM subchannel starting at 5.75 MHz above the channel lower limit (4.5 MHz above the video carrier) with a maximum deviation of ± 25 kHz. The video signal also includes provision for various control signals related to picture synchronization.

A simplified block diagram of a television receiver is shown in **Fig 10-5**. Note that it starts out the same as an AM receiver with a *video detector* demodulating the spectrum of Fig 10-4. The AM video signal is applied to the picture tube to control video brightness and color information. At the same time, the vertical and horizontal synchronization information is extracted and used to synchronize vertical and horizontal oscillators and amplifiers. These are used to drive the cathode-ray tube (CRT) deflection coils, to move the beam back and forth and up and down the screen.

At the output of the video detector, the audio carrier and sidebands appear as if they were just a 4.5-MHz FM signal. A standard FM receiver, set for 25-kHz deviation is used to reproduce the audio signal.

One fact that is not apparent from what I've said is the frequency range required. In the days when TV signals were mostly sent by radio, the receivers first just received the VHF band. This consisted of two segments: 54 to 88 MHz (channels 2 to 6, with a gap from 68 to 76 MHz) and 174 to 216 MHz (channels 7 to 13). Later the UHF channels were added, originally 470 to 890 MHz (channels 14 to 83) and later, 470 to 806 MHz, to provide for cellular telephone service in the upper portion of the original TV UHF band. This provided a challenge to receiver designers of the day—to design radios that could cover a frequency range from lower VHF to UHF, almost into the microwave region. And yet these receivers had to be easily (and at low cost) manufactured and used by the general public. The migration to cable systems with hundreds of channels just made things even more challenging, with a frequency range from HF to SHF. The transition from vacuum-tube to solid-state technology in the sixties made it all feasible.

As this is being written, a migra-

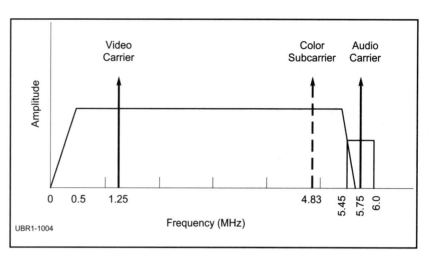

Fig 10-4—Spectrum definition of the North American standard analog television signal.

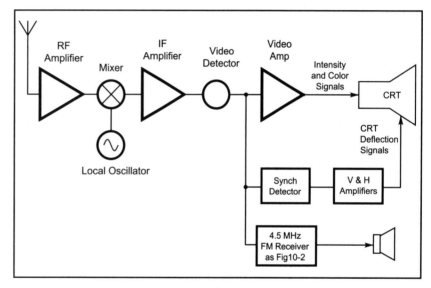

Fig 10-5—Simplified block diagram of an analog television receiver.

tion is occurring towards digital television transmission. Standards are becoming established, but it's clear that the analog baseband signal we have been discussing will be replaced with a pulsed digital format.

Pulsed Radar Systems

Radar is an interesting application of radio that has a number of differences from most communication systems we have discussed to this point—yet the underlying technology is very similar. Most (but not all) radar systems have the transmitter and receiver in the same location and they share the same antenna. Most radar systems convey no explicit information in their signal—truly a communications system in which the medium *is* the message.[1] On the other hand, the transmitter sends out a signal, rather like a Morse code dot, and the receiver receives the same—over and over. The radio portion of the system is just like any other in many ways.

The basic operation is easy to understand. The transmitter at regular intervals (the *Pulse Repetition Interval*, or PRI) sends a pulsed radio signal towards the sky. If nothing comes back, there is no signal to receive. If an aircraft is in the path of the signal, it will reflect back a small bit of the signal (called an *echo*) that can be detected. You can measure the elapsed time from transmission to reception and, since you know the speed of light, you can calculate the distance to the target. If the antenna has a narrow beam and is rotating, you can also determine the direction to the target.

The radar sends its pulses under conditions that are well established and understood. If you want the radar to observe a particular range, you need to have an interval between our pulses at least long enough for the signal to get that far and return. A typical example is a ground-based radar designed to detect and track aircraft for air traffic control or military applications. A curved earth and limited flying altitude establish the maximum range of interest as shown in **Fig 10-6**. This results in a round-trip propagation time of about 3 msec, a fairly standard pulse repetition interval aircraft (PRI) for search radar and perhaps the first fundamental radar parameter.

The second fundamental radar parameter is the signal *pulse width*. A wide pulse can be easily received in a receiver with a relatively narrow bandwidth, however, all targets that are encountered within the pulse time will tend to appear as a single (although wider) reflection signal rather than as distinct targets. This is obviously a serious problem for both the air-traffic control application (it would be great to know if there were one or two aircraft in that little piece of sky) as well as for the military (how many interceptors do we need to scramble?). This parameter is referred to as *range resolution* and is the distance that the pulse travels during the length of the pulse. A typical pulse width used is 1 μsec. The distance the signal travels during the pulse interval is about 1000 feet, a reasonable resolution for most purposes.

Fig 10-7 shows a view of what would be seen on an oscilloscope with such a system. This was the earliest type of radar display (WW II era) and was called an *A-scope*. The

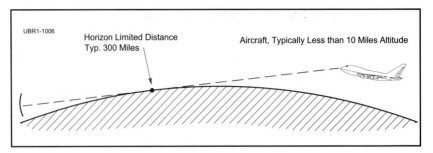

Fig 10-6—Operating environment for typical ground-based search radar.

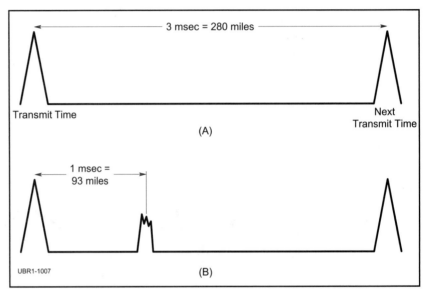

Fig 10-7—A search radar signal without targets (at A) and with targets (at B).

A-scope provided a view of one slice of the environment corresponding to the antenna beamwidth. The more commonly seen circular scope with the beam synchronized to the antenna position to provide azimuth information was called a *Plan Position Indicator* (PPI) or *B-scope*. A target's distance at a particular PRI can be easily read from the scope display by appropriate calibration.

A major difference between data communications and radar radio equipment is that the data signal tends to be on about as much time as it's off, for about a 50% transmitter *duty cycle*. This depends on the coding[2] and can be as much as 100% for frequency-shift keying, for example. On the other hand, a radar system is transmitting only for 1 μsec every 3 msec, a duty cycle of

about 0.03%. This has an impact on transmitter and power supply design, in particular.

Another radar parameter is the transmit power required to receive a detectable return echo. While this is an issue in all radio systems, we will show in Chapter 11 that a radio system's received power is reduced by the square root of the distance, while a radar system's signal level drops with the fourth power of distance, since the transmitted pulses must travel to the target and then travel back to the receiver. Long-range radars need *lots* of peak power.

Notes
[1]With apologies to Marshall McLuhan.
[2]In fact much more clever encoding schemes are employed to improve the average signal-to-noise ratio.

Review Questions

1. Discuss reason why the US long-distance telephone network found it desirable to move from analog to digital multiplexing.

2. What is the average transmitter power (over time) if a radar with 1.0 μsec pulse width and a 3 msec Pulse Repetition Interval has a peak power output of 10,000 W?

3. Consider the appearance of a radar return from a target beyond the design maximum range of a radar system. How would it compare to a target within the range?

Chapter 11

Superheterodyne Receivers —Making Them Sing

He's loud for a good reason! Andi, UA3DPX, of Russia does very well in HF DX contests with this antenna setup.

Contents

There are a few fundamental design considerations I should mention up front so you don't leave the previous chapter thinking that the basic superhet is without limitations of its own. First, you should be aware of the range of signal amplitudes that a receiver should be able to listen to. The determination of required receiver sensitivity and management of receiver gain are key considerations in effective receiver design. Next, I will discuss some signals you don't ever want to listen to—image signals and intermodulation products. Both involve reception of undesired signals that must be taken into account.

RECEIVER AMPLITUDE RESPONSE

When I discussed crystal sets in Chapter 1 and 2, I noted that to increase the loudness of the signal you were listening to, you had to either install a larger antenna or move closer to the transmitter. By adding audio and RF amplifiers, you could compensate for those shortcomings to a certain extent. But it's easy to fall into the trap of thinking that you can get as loud a signal as you'd like just by adding enough amplifiers.

There are limits to how far you can go in this direction. If you amplify enough, you will find that in addition to amplifying the desired signal, you start to amplify noise. This is ultimately the factor that limits our ability to receive signals. In general, it's fair to say that noise defines the lower bound of signals that you can receive. In fact, if the signals are much lower than the noise, you can't hear the desired signals at all. If the signals are stronger than the noise, you have a chance of doing something with them. Some approximate reference points may give an idea:
- A highly trained Morse operator can actually decode Morse signals that are somewhat weaker than the noise.

- With signals about 100 to 1000 times the noise power (giving a 20 to 30 dB signal-to-noise ratio, SNR), automated systems can make a bit/no-bit decision and be correct most of the time (not always, because noise power is specified as the average level, so sometimes a noise peak will be much higher than average).
- A telephone call with a 30 dB (1000 times power) SNR is considered "toll quality."
- A television signal with a 50 dB SNR will have virtually no visible "snow" in the background.

Where Does Noise Come From?

There are a number of answers to this one, because there are multiple noise sources to consider. While all are always present, one or more of the following sources are usually dominant, depending on the frequency or other parameters.
- Atmospheric Noise. This is noise generated within our atmosphere due to natural phenomena. The principal cause is lightning, acting very much like an old-fashioned gigantic spark transmitter and sending wideband signals great distances. All points on the Earth receive this noise, however, it is much stronger in some geographic areas than others, depending on the amount of local lightning activity. Lightning is usually the strongest noise source in the LF range and may dominate well into the HF region, depending on the other noises in the region. The level of atmospheric noise tends to drop off by around 50 dB every time the frequency is increased by a factor of 10. This source usually drops in importance by the top of the HF range.
- Man-made noise. This source acts in a similar manner to atmospheric noise, although it is more depen-

dent on local activity rather than on geography and weather. The sources tend to be rotating electrical machinery, as well as gasoline engine ignition systems and some types of lighting. All things being equal, this source drops off on average by about 20 dB every time the frequency is increased by a factor of ten. The higher frequency is due to some sparks having faster rise times than lightning. The effect tends to be comparable to atmospheric noise in the broadcast band, less at lower frequencies and a bit more at HF. Your mileage may vary!
- Galactic Noise. This is noise generated by the radiation from heavenly bodies outside our atmosphere. Of course, while this is noise to *communicators*, it is the desired signal for *radio astronomers*. This noise source is a major factor at VHF and UHF and is quite dependent on exactly where you point an antenna (antennas for those ranges tend to be small and are often *pointable*). It also happens that the earth turns and sometimes moves an antenna into a noisy area. If the sun, not surprisingly the strongest noise signal in our solar system, appears behind a communications satellite, communications are generally disrupted until the sun is out of the antenna's receiving pattern.
- Thermal Noise. Unlike the other noise sources, this one originates in our equipment. All atomic particles have electrons that move within their structures. This motion results in very small currents that generate small amounts of wideband radiation. While each particle's radiation is small, the cumulative effect of all particles becomes significant as the previous types of noise sources roll off with increasing frequency. The reason that this

effect is called *thermal* noise is because the electron motion increases with the particle's temperature. In fact the noise strength is directly proportional to a change in temperature, if measured in terms of absolute zero (0 degrees kelvin, abbreviated 0 K). For example, if we increase the temperature from 270 to 280 K, that represents an increase in noise power of 10/270 = 0.037, or about 0.16 dB.

Why do we Care?

A receiver designer needs to know how strong the signals are so you can establish the range of signals the receiver will be required to handle. Since noise power is a widely varying average, it is a good idea to set the sensitivity floor at perhaps 10 to 20 dB below the expected noise. As noted above, this is related closely to the frequency we want to receive. **Table 11-1** below is a rough guide of the noise power you might expect to deal with in each frequency range. Note that for frequencies at which external noise sources are most important, the noise power (and signal power) will also be a function of the antenna design. In other words, pick up more signals with a more effective antenna and you also generally get more noise!

Table 11-1 gives us an idea of how small a signal we might need to deal with. A designer must create a receiver that will handle signals from below the noise floor to as strong a signal as the closest nearby transmitter can generate. Most receivers have a specified (or sometimes not) highest input power that can be tolerated, representing the high end of the spectrum. Usually the maximum power specified is the power at which the receiver will survive without being damaged (see **Fig 11-1** for an example of a mid 1950s deluxe receiver, the Collins 75A4). A somewhat lower power level is generally the highest signal that the receiver can receive without overload.

Table 11-1
Typical noise levels (into the receiver) and their source, by frequency.

Frequency range	Dominant noise sources	Typical level (dBm)[1]*
LF 30 to 300 kHz	atmospheric	−16.6
MF 300 to 3000 kHz	atmospheric/man-made	−39.7
Low HF 3 to 10 MHz	man-made/atmospheric	−31.7
High HF 10 to 30 MHz	man-made/thermal	−47.3
VHF 30 to 300 MHz	thermal/galactic	−96.6
UHF 300 to 3000 MHz	galactic/ thermal	−136.6

*The level assumes a one square meter antenna size for LF, MF, VHF and UHF. A half-wavelength antenna is assumed at HF.

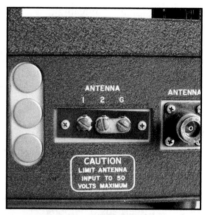

Fig 11-1—The warning on this receiver is clear about the maximum voltage that can be tolerated coming from the antenna. *(Photo courtesy Chris Galbraith, KA8WFC.)*

The highest signal you will want to receive is very much dependent on the application and the operational environment. For a standard broadcast receiver, you can easily imagine living next door to a transmitting antenna, and you would expect to be able to receive that signal, as well as signals from distant stations. A point-to-point microwave system will likely have a much smaller variation in signal strength, since it will receive a signal from a single transmitter, at a fixed distance from you. A naval vessel or military headquarters may have multiple communications systems running simultaneously, and will need to receive weak signals while strong signals are on nearby frequencies (see also the next few topics). An airport search radar receiver will need to deal with reflected signals from aircraft 200 miles away as well as reflections from nearby structures or aircraft, perhaps only 500 feet away.

The range of signals a receiver must respond to can run from less than 0.1 pW (pico watt) to as much as 1 W. This range defines the *dynamic range* (the difference between strongest and weakest signals) of 130 dB. This is a very large range, and it is unlikely that any

receiver can deal with the full range without somehow adjusting the gain of some of its stages. Even if the receiver could work over such a range, the volume would be more than our ears would like to hear!

How do You Handle Such a Range of Signals?

A first reaction might be to set all gains in receiver amplifiers to allow reception of the weakest signal, and then reduce the *volume* or gain of the audio amplifier to provide the desired sound level for any signal in the range. While this is a logical conclusion, and would work if all the amplifiers were linear over the whole range, unfortunately they aren't! If this approach were followed when very strong signals are received, it is likely that RF or IF amplifiers would be overloaded and would clip and distort the signal. Even if the volume were turned down, the resulting signal would not be satisfactory because of the distortion.

The locations at which receiver gain could be adjusted are shown in **Fig 11-2**. Some receivers make use of all of them, while others use fewer. The opposite end of the receiver from the audio gain proposed below is the location for the adjustable RF attenuator shown on

the left of the figure. The RF attenuator is the best approach to avoid overload problems associated with just adjusting the audio gain. It can be very effective, since it can lower the receiver input to any desired level. Unfortunately, as the input is reduced, the signal gets progressively weaker while the thermal noise in the receiver stays the same. If too much attenuation is used, you will not hear the desired signal because of the noise of all the amplifiers. Note that since each amplifier amplifies both the desired signal and the noise from all stages in front of it, the first amplifier (the RF amplifier) often is the biggest contributor to internal thermal noise.

Automatic Gain Control— Why Didn't I Think of That?

In addition to the problems of overload, distortion and reducing signal-to-noise ratio; early radio operators grew tired of lunging for a manual gain control whenever they tuned across a signal that was much louder than the one they had been listening to. Before they all lost their hearing, someone invented *automatic gain control* (AGC), in earlier times known as *automatic volume control* (AVC). This great idea makes use of a sample of the average level of

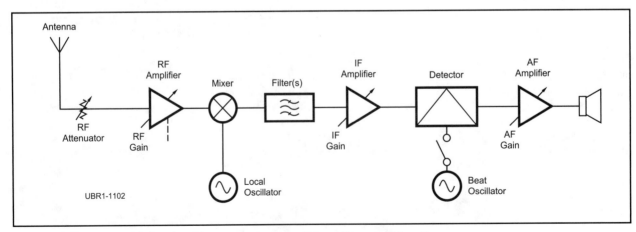

Fig 11-2—A superheterodyne receiver, showing possible gain-control locations.

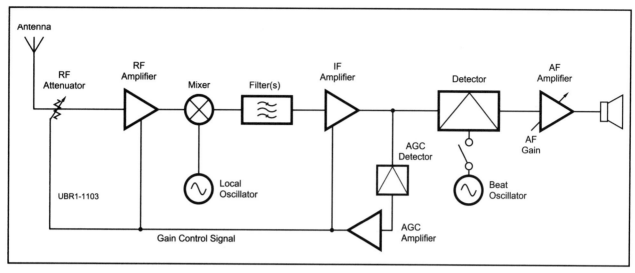

Fig 11-3—A superheterodyne receiver with automatic gain control (AGC).

received signal to control the gain of the receiver so that the output stays constant over most of the dynamic range of the receiver. Most users of radio equipment take this for granted and don't even know that it is happening.

Our block diagram of Fig 11-2 has been enhanced with the inclusion of AGC in **Fig 11-3**. Note that while it's easy to describe, as above, and easy to show in a simple diagram, the design of an AGC system must be performed carefully to insure that the receiver's signal-to-noise ratio is optimized at all gain levels. This means, for example, that the RF gain should be held high until the level gets close to the overload point. *Delayed AGC* is a term that is used to describe an AGC system that makes no gain adjustment at the front end of the receiver for very weak signals, delaying the effect until they are well above the receiver's noise floor.

With a well-designed AGC system, the signal level entering the detector is set to the optimum level for any strength within the receiver's dynamic range. This means that no stage ahead of the detector can overload and the audio gain, or volume control, can be set to whatever sound level the listener prefers. This is why even the usual "kitchen radio" just has a tuning knob and a volume control. Everything else happens "under the covers!"

AGC for Different Signals

The earliest AGC was used for the reception of AM broadcasts—that was the early kitchen radio, before FM was popular. For AM operation AGC is fairly straightforward. Another detector is provided that has a long time constant and responds to the average level of the received carrier. The higher the level, the less gain the receiver provided. In the early days of "modern" radio (say through the 1950s) that was the only type of signal that AGC had to deal with. As more and more services migrated to SSB, there was a need to provide a similar function for those signals, and AGC detectors were adapted to take longer averaging times to adjust the gain for signals that didn't have a constant carrier level. Some modern receivers have adjustable attack and decay times to allow optimization for different signals and conditions. In all cases, the principle is the same—measure the level, compute the average as appropriate and adjust the gain.

Gain Control in Other Systems

The kind of AGC we have discussed is found in most types of radio equipment—FM radios, televisions, microwave links, etc. This shouldn't be too surprising, since all must deal with signals of multiple levels and none do well if overloaded. One type of system that operates a bit differently is a radar receiver. The radar receives signal of widely different amplitude, but unlike communications systems, the radar system obtains information about the size of the target based on received signal strength, so gain adjustment needs to be handled differently to avoid loss of information. In a radar system, the receiver can predict *when* the signals will be very strong.

Following the transmission of an outgoing pulse, the radar will first receive reflected signals from its immediate surroundings, then from aircraft nearby and later from more distant aircraft. In Chapter 23 we will show that, unlike communications systems in which the received signal is reduced by the square of the distance, with radar the received signal drops off with the *fourth* power of distance. Distant reflections are thus much weaker than nearby ones, and the highly sensitive receiver generally does better if the gain is reduced for nearby signals.

For the case of our long-range

search radar with a 3 msec pulse repetition rate, closer signals will be received within the first msec after the pulse is transmitted. A system known as *sensitivity time control* (STC) operates much like AGC, but is based on the time since the start of the transmit pulse. With radar, the amplitude of the reflected signal can provide a measure of the size of the target and it is not desired to make all signals seem the same level, as is done with AGC. **Fig 11-4** shows how the gain might be adjusted in an STC equipped radar receiver.

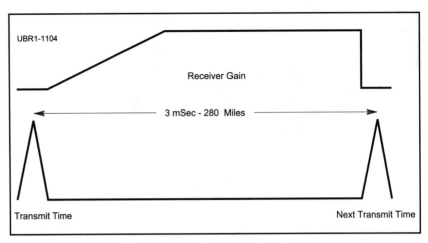

Fig 11-4—A search-radar with sensitivity time control (STC).

Superhet Image Response

In addition to dealing with the signals we want to receive, with a superhet we also need to deal with a number of types of signals we *don't* want to receive at all. When we discussed mixers, we noted that a mixer would provide an output into the IF amplifier of signals at the local oscillator *plus* the IF frequency, and also at the local oscillator *minus* the IF frequency. Let's do an example to make this clear. Let's say our IF frequency is 455 kHz and we want to receive a signal with a frequency of 14,000 kHz. We could set our local oscillator (LO) at 14,455 kHz and we would translate the 14,000 kHz signal to 455 kHz by accepting the difference frequency. All is well and good.

Then the Image Appears

Let's consider the case in which we also have a signal that we *don't* want to hear at 14,910 kHz. Our receiver would also respond to that signal since 14,455 kHz *plus* our IF of 455 kHz is 14,910 kHz. Thus, unless we take measures to avoid it, we will also translate this signal to 455 kHz IF. This is called an *image* signal. We now have to deal with two signals at the input of our receiver, when we wanted just one.

Image response is the result of the fact that a superhet will respond to signals above and below the LO signal unless we eliminate the undesired signal before it reaches the mixer. The answer is to have the RF stage tuning sufficiently sharp to respond to 14,000 but not to 14,910 kHz. Note that this is much easier than in a TRF receiver, in which the RF stages must eliminate the next channel, perhaps at 14010 kHz.

It is important to take this effect into account even if there is no station at 14,910 kHz. Any atmospheric or other noise at 14,910 kHz will be translated to 455 kHz even if there is no signal directly on that frequency. The noise received will likely double the noise power, while the desired signal level remains unchanged. This results in a reduced signal to noise ratio.

Receivers covering the standard AM broadcast band (550 to 1750 kHz) and using a 455 kHz IF can often provide sufficient image rejection with a single tuned circuit and thus do not even have an RF stage in front of the mixer. For higher frequencies, up to around 15 MHz, a single RF amplifier with two or three tuned circuits is usually sufficient to provide adequate image rejection for most applications.

Note that the higher the IF frequency, the further away the image frequency is. And conversely, the lower the IF, the closer is the image—making it harder to eliminate. On the other hand, using regular LC circuits to create selectivity, the higher the IF, the wider the selectivity becomes. The lower the IF, the sharper the selectivity, which also means the bandwidth is narrowed. So you have yet another design trade-off! The broadcast radio standard of 455 kHz is a compromise that works well for AM broadcast radios and receivers into the lower HF region.

Early communications receivers using the superhet architecture addressed this problem by adding RF stages with tuned circuits on each side to amplify the desired signal while attenuating the image, located 910 kHz away. This required large ganged multi-section tuning capacitors with associated alignment efforts to achieve the desired result. This was the state of the communications receiver art from the mid 1930s to the mid 1950s.

Double Conversion— Another Good Idea

A different technique, *double conversion*, was created to solve the image problem. Rather than having to decide between a high IF frequency for good image rejection, or a low IF frequency for narrow channel selectivity, some bright soul decided to do *both*! As shown in **Fig 11-5**, a conversion of the desired signal to a relatively high IF is followed by a conversion to a lower IF to set the selectivity. This arrangement solved the image problem nicely, while the rest of the receiver could be pretty much kept the same as before. A number of manufacturers made receivers in the 1950s that added an additional conversion stage, for reception above 10 or 15 MHz, where the image response had been poor with single-conversion radios.

Some manufacturers decided that if two conversions were good, three must be even better. These folk started in the same place, but converted the 455 kHz second IF down to 50 or 100 kHz for even sharper selectivity. This approach lasted until crystal lattice filters became available that outperformed the low frequency LC circuits.

Shortly after WWII a visionary radio pioneer, Arthur Collins, founder of Collins Radio in the 1930s, came up with another approach to double conversion. One problem with earlier MF and HF receivers was that they used a standard LC (first) oscillator using a variable capacitor of the type used in broadcast receivers. As noted earlier, these capacitors typically had a 9:1 capacitance range resulting in a 3:1 frequency range. The typical tuning arrangement for multi-band receivers was 0.5 to 1.5, 1.5 to 4.5, 4.5 to 13.5, and 13.5 to 30 MHz. This covered from the MF AM broadcast band to the top of the HF range in four bands.[2] With this arrangement, the tuning rate and dial calibration marks got less and less precise as higher bands were selected, making

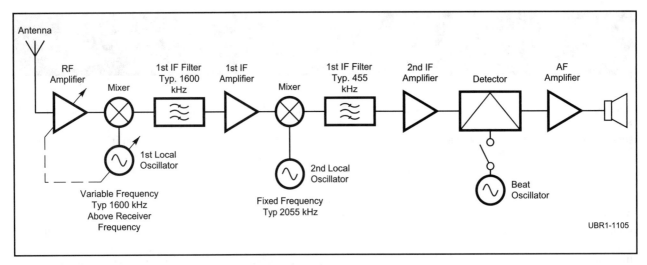

Fig 11-5—Double-conversion superhet receiver. Early type.

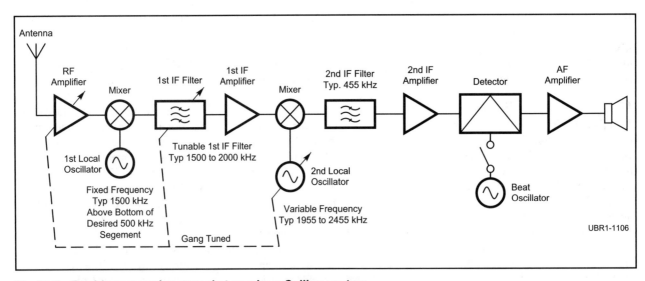

Fig 11-6—Double-conversion superhet receiver: Collins system.

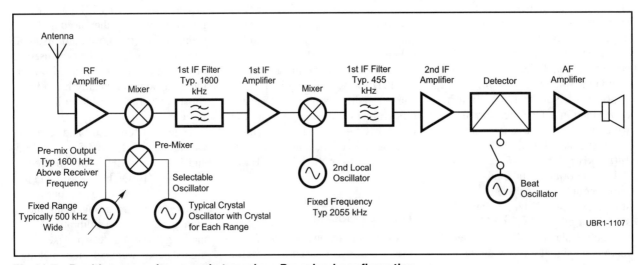

Fig 11-7—Double-conversion superhet receiver. Pre-mixed configuration.

exact tuning difficult. The Collins system (see **Fig 11-6**) moved the variable oscillator to the second mixer and used crystal-controlled oscillators as the first oscillator. Although many more bands were required (30 for the famous 51J series), each tuned with exactly the same tuning rate—that is, the rate set by the second oscillator.

Collins went a step further and designed an inductance-tuned oscillator that was linear throughout its range. These oscillators could be tuned to the nearest 1 kHz from beginning to end, thus avoiding the tuning uncertainty of other receiver types. For this to work properly, synchronized or *gang* tuning is required between the oscillator, first IF variable filter and RF stages. This was no problem for the engineers at Collins Radio, but their equipment brought a premium price.

A third approach to double conversion was a mix of the first two (no pun intended). The *pre-mixed* arrangement, **see Fig 11-7**, uses a single variable oscillator range, like the Collins, but does the mixing outside of the signal path. This avoided the need for variable tuning at the first IF, allowing a tighter *roofing filter* (more about this in the next section) but traded it for the need for filtering at the output of the pre-mixer (not shown).

Back to the Future—Single Conversion Returns!

Receivers of today can be found in any of the previous configurations. In addition, over the last quarter century there have been remarkable breakthroughs in the development of crystal lattice filters at progressively higher frequencies. For years, reasonably priced crystal filters have been available throughout the HF region. This allows a single conversion receiver to obtain the desired selectivity characteristics, while still having a high enough IF frequency to avoid image problems. Receivers that cover a wide tuning range, however, often have a first IF frequency in the VHF region. Crystal filters have become available well into the VHF region.

The Image-Rejecting Mixer—What a Concept

In our discussion of types of mixers in Chapter 6, we discussed the image-canceling mixer and showed the signal relationships in Fig 6-6. This kind of mixer is used to reduce image response and has been used for many years in fixed frequency systems (radar for example), but not often for variable frequency receivers since it can be difficult to maintain the phase relationships as frequency is changed. Some recent receivers have taken advantage of this technique, often in addition to dual conversion, to improve image rejection still further.

Intermodulation Distortion

Another type of undesired signal is a direct result of the mixing process, although it also happens in amplifiers, particularly as they approach their amplitude limits and start to saturate. *Intermodulation distortion* results from the fact that no real mixer has a perfect square-law response. If its response curve is not exactly $y = x^2$, it can be expressed mathematically as the infinite series $y = k_0 + k_1x + k_2x^2 + k_3x^3 + k_4x^4 + \ldots + k_nx^n$. This expression can be used to describe any two-dimensional curve.

Note that if all the k coefficients above k_1 equal zero, it describes a straight line. If only k_0 and k_1 are non-zero, it describes our ideal mixer's parabolic response curve. The higher order terms result in other combinations of two frequencies. For real mixers, the terms above the x^2 term fall off rapidly and the x^3 term causes the most problems. The so-called *third-order response* to two signals can be described as follows: $(a + b)^3 = a^3 + 3ab^2 + 3a^2b + b^3$.

The terms $3ab^2$ and $3a^2b$ represent product terms of a sinusoid at frequency a, and another at the frequency 2b, for the first term, and the other way around for the second term. In a mixer, you get an output at the sum and difference. So considering the first term, you get an output from 2b–a.

In this case, the a and b signals are signals you don't want to hear. These are outside the receiver's IF bandwidth but can get into the mixer. Remember that a squared term has a twice frequency component. Let's say you are trying to listen to a station at 1560 kHz and that there is a strong station 20 kHz above this frequency (1580 kHz is b in the above discussion) and another 40 kHz above a, at 1600 kHz. Note that if the IF stages have a bandwidth of 15 kHz (±7.5 kHz), you wouldn't hear these stations if the mixer were a perfect square law device.

Because there is a non-zero coefficient for the x^3 term, you will get the response of *a* times twice *b* resulting in the sum and difference frequencies. Note that the difference between 1600 and 2 ×1580 = 1560 kHz. Thus the imperfect mixer translates two signals you shouldn't be able to hear at all into a new signal right on top of the one you are trying to listen to!

This effect is called *third order intermodulation distortion* (3OIMD) and can be a serious limitation of a superheterodyne that tries to operate in the presence of many strong signals. There are two ways to make this effect manageable, and good receivers employ both. The first is to make the mixer as close to a square law device as possible, with a very small third order coefficient. The second approach is to put a filter in front of the mixer to eliminate signals that are further from the desired signal than needed to just hear the desired signal. The best receivers can provide up to about 90 dB of reduction of 3OIMD signals.

Notes

[1] *Reference Data for Radio Engineers, Fourth Edition*, International Telephone and Telegraph Corporation, 1956, New York, NY, p 763. Data derived assuming a half-wavelength antenna in mid band.

[2] A notable exception was the Hammarlund Manufacturing Corporation whose receivers used a 2:1 tuning range. The typical bands were 0.5 to 1, 1 to 2, 2 to 4, 4 to 8, 8 to 16 and 16 to 32 MHz. By covering the range in six rather than four bands, the tuning was noticeably more precise, especially at the higher bands.

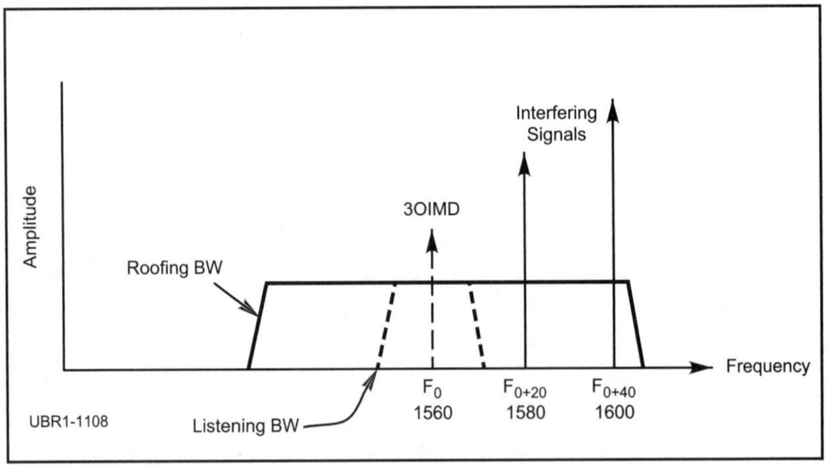

Fig 11-8—Graphical representation of two-tone third-order IMD.

1. Imagine an AM car radio if AGC were not part of the design. What would you have to do as you drove along listening to the radio?

2. If you have a superhet with the local oscillator tuned to 25 MHz and an IF of 455 kHz, what frequency signals could you receive? Repeat the exercise for a 3 MHz IF. Discuss the differences you observe.

3. If you used LC circuits for IF selectivity in Question 2, discuss the difference in performance between the two IF frequencies.

4. You are trying to receive a weak signal on a frequency of 15 MHz. There are strong signals at 15.007 and 15.014 MHz. Describe the problem you might have listening to the desired signal.

Chapter 12

Transmitters—the Other Side of the Equation

Bill Stein, KC6T, built this elegant homemade multiband Yagi beam antenna.

Contents

So far, I have been talking mostly about receivers, with the implicit assumption that there must be some transmitters out there to which you can listen. Now I will switch over to talking about transmitters. Fortunately, much that we have discussed about receivers will help you understand how transmitters are put together, since they both are made up of the essentially the same pieces.

While transmitters have many of the same building blocks used in receivers, it's important to keep in mind that the components used may not be the same size. An RF amplifier in a receiver may deal with amplifying picowatts, while one in a transmitter may output up to megawatts. While the circuits may even look similar, the size of the components—especially cooling systems and power supplies—may differ significantly in scale. We will discuss RF power amplifiers and high-powered transmitters in more detail in the next chapter.

A Simple Data, Telegraph or Pulse Transmitter

The very simplest transmitter consists of an oscillator that generates a signal at the desired transmitting frequency. If the oscillator is connected to an antenna, the signal will propagate outward and be picked up by any receivers within range.[1] Such a transmitter will carry little information, except perhaps for its location. This could serve as a rudimentary *beacon*, for direction finding or radiolocation, although real beacons transmit identification data. A true beacon also indicates whether or not it is turned on, perhaps useful as part of an alarm system.

To actually transmit information, you must *modulate* a transmitter. The modulation process was introduced in Chapter 2 and involves changing one or more of the signal parameters

to apply the information content. This must be done in such a way that the information can be extracted at the receiver. As previously noted, the parameters available for modulation are:
- *Frequency*—This is the number of cycles the signal makes in a second's worth of time.
- *Amplitude*—Although the amplitude, or strength, of a sinusoidal signal is constantly changing with time, we can express the amplitude by the maximum value that it reaches.
- *Phase*—The phase of a sinusoidal signal is a measure of when that sinusoid starts, compared to another sinusoid at the same frequency.

You could use any of the above parameters to modulate a simple transmitter; however, the easiest to visualize is probably amplitude modulation. If you were to just turn the transmitter on and off (with on standing for a binary "one" and off for a binary "zero"), you could surely send data. Unfortunately,

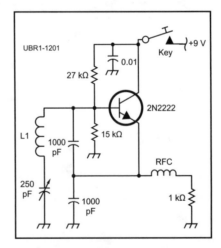

Fig 12-1—A simple solid-state, low-power HF oscillator-transmitter built by Chuck Counselman, W1HIS. L is a pre-assembled coil stock; 3-inch diameter, 6 inches long and 4 turns per inch.

some care is needed in how you implement such a function. Let's say that you perform the obvious step of just removing and turning on the power supply. You might be surprised to find that it takes too much time for the voltage to rise sufficiently at the oscillator to actually turn it on at the time you make the connection.

Similarly, you might be surprised to find that when we turn off the power you would still be transmitting for some time after the switch is turned off. These finite intervals are referred to as *rise* and *fall times*. They generally depend on the value of filter capacitors in the circuit—designed to remove extraneous signals, ac hum and noise from the power wiring.

Here is an easy to visualize analogy—the use of light to transmit code signals. If you've ever seen a WW II movie showing a Navy Morse operator sending data between ships with a lantern, you may have wondered why they used a special mechanical shutter device instead of turning the light bulb on and off.[2] One reason is that the bulb continues to radiate for a period while the filament is still hot, even after the power goes off.

A simple, low power HF Morse oscillator transmitter is shown schematically in **Fig 12-1** with a photo of an example in **Fig 12-2**.[3] This transmitter will generate a signal in the MF broadcast band and can be heard at short distances without any antenna.[4]

Fig 12-3 is the schematic of a crystal oscillator transmitter designed for HF operation. By using a crystal-controlled oscillator, you gain frequency accuracy and stability, but lose a certain amount of flexibility because the frequency is determined in largest part by the mechanical properties of the crystal itself.

In this design the key just turns the

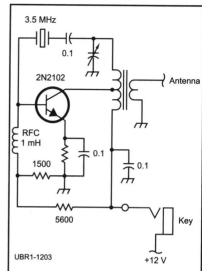

Fig 12-3—A simple solid-state low-power HF crystal controlled oscillator-transmitter. The unspecified tuned circuit values are resonant at the crystal frequency.

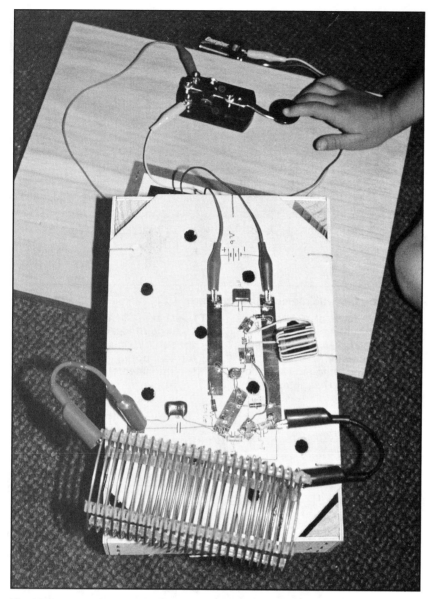

Fig 12-2—Photo of the simple solid-state, oscillator-transmitter. *(Courtesy Chuck Counselman, W1HIS.)*

power on and off. The 0.01 µF line bypass capacitor was chosen so that it would not delay the rise or fall at reasonable keying speeds. This transmitter actually works and can generate a few mW of RF power at the crystal frequency. However, it has a number of limitations. These limitations are in common with the early vacuum-tube oscillator transmitters of the 1920s—although they were a major improvement over the spark transmitters that preceded them. First, the oscillator is dependent on the environment. Changes in the antenna, due to wind, for

example, could change the load and cause the frequency to shift very slightly. Second, the oscillator generates not only a signal at the crystal's fundamental frequency, but also at harmonics (multiples) of the frequency. Third, every time the key is depressed, the oscillator must start up, possibly taking a finite time to stabilize at its rest frequency. Since the frequency could change slightly due to the first and third factors, the sound of such a simple transmitter often might have a "chirp" or even a "yoop" to it. Imagine the "yooping" sound generated by a bird warbling

at the keying rate of a Morse Code transmission!

The Next Step—a Two-Stage HF Transmitter

By adding an amplifier stage to your little transmitter, you can address the first problem of frequency change while the oscillator starts up. While we're at it, you should add filtering to solve the second problem of harmonics. I will defer dealing with the third problem, but you could deal with this too if you needed to, by not keying the crystal oscillator itself, only keying the following amplifier stage(s).

Fig 12-4 is the circuit of a much better, but still simple radiotelegraph transmitter, called the *Tuna Tin 2* because it was built to fit in a tin can that once contained tuna fish. This circuit uses an oscillator similar to that of Fig 12-3, but followed by a stage of amplification. The amplifier increases the output power to about $1/3$ W. Note that the output circuit includes a low-pass filter to reduce any harmonic output below the levels required by the US Federal Communications Commission (FCC).

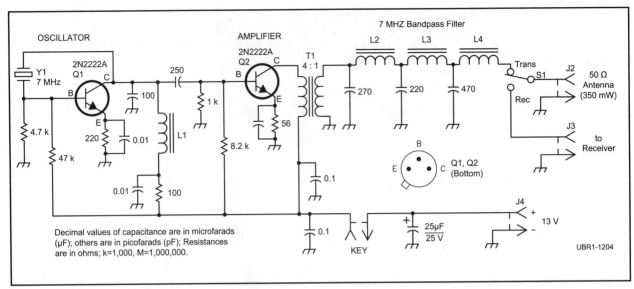

Fig 12-4—Circuit diagram of the *Tuna Tin 2*, two-stage radiotelegraph transmitter.

J1—Single-hole-mount phono jack. Must be insulated from ground. Mounts on printed circuit board.

L1—22 µH molded inductor.

L2—19 turns of #26 wire on a T-37-2 toroidal core.

L3, L4—21 turns of #24 wire on a T-37-6 toroidal core.

Q1, Q2—2N2222A or equivalent NPN transistor.

S1—Antenna changeover switch. Miniature SPDT toggle.

T1—4:1 broadband transformer. 16 turns of #26 wire on the secondary, on an FT-37-43 toroidal core.

Y1—Fundamental crystal, 7 MHz.

A simple single-stage oscillator/transmitter can be useful, and is still employed in special circumstances, particularly in the microwave region. However, one advantage of separating the oscillator from the device providing the output power is that, for applications requiring frequency stability, an oscillator that doesn't have to deliver much power is more easily stabilized. The configuration is sometimes referred to as a *master oscillator-power amplifier* or MOPA. (This signifies that the oscillator is master of the frequency and the power amplifier delivers the needed output.)

While ¹/₃ W doesn't sound like a lot of power (and it isn't!), this transmitter has been built by many Amateur Radio operators and used for worldwide communication on the 7-MHz (40-meter) amateur band. **Fig 12-5** shows the original model fitted in its tuna fish can.[5] If you have an Amateur Radio operator's license and are authorized to operate in this band using radiotelegraph, you may

Fig 12-5—Photo of the original *Tuna Tin 2*, built into a tuna fish can.

want to build one of these and try your hand at low-power operation. You will be surprised at how effective it is, so long as there isn't a stronger signal on your frequency!

Moving Beyond the Tuna Tin

The *Tuna Tin 2* is a real radio transmitter and has been used for worldwide communication. The output power of ¹/₃ W can be a limitation, however. Most transmitters provide a higher output power, especially on HF. The *Tuna Tin 2* transmitter does have all the essential elements, however, and is a good example of a radiotelegraph transmitter design.

The low ¹/₃ W level requires considerable operator skill, effective antennas, very good propagation and, yes, a bit of luck, to get through. While you may be able to talk somewhere most of the time, you may not be able to get your signal where you want it. In the next chapter, we will discuss power amplifiers for transmitters. Additional amplifier stages may be added to this (or any other simple transmitter) as needed, to achieve whatever output power is required.

This is where the decibel earns its keep! A power output of ¹/₃ W is about –5 dBW, or 5 decibels below 1 W. A typical amplifier stage might have 15 dB of gain. An additional stage at that gain would provide an output of –5 dBW + 15 dB = 10 dBW or 10 W. Another stage with 15 dB of gain would yield 25 dBW, which is 316 W. Another stage would bring this to 40 dBW or 10,000 W. Not very many stages and before long we're talking serious power here! Although higher-power amplifiers are bigger, more expensive and tougher to build (not to mention noisier when they blow up), they are conceptually similar in design.

Transmit-Receive Switching Control

The careful observer of Fig 12-4 will note that the transmitter output is routed through a *transmit-receive* switch S1. Some kind of transmit-

receive switching is required for most two-way radio systems, and it's worth digressing for a moment since the switching may be an important consideration in the design of the transmitter.

Note that different types of systems will have different switching requirements. There are generally three broad categories of transmit-receive switching:

- *Simplex operation*—This describes one-way transmission. Typically doorbells and broadcast transmitters operate as simplex systems.
- *Half-duplex operation*—This describes a two-way system that can operate in either direction, but only one-way at a time. A taxi radio with a push-to-talk button on the microphone is a half-duplex system, as is the *Tuna Tin 2*.
- *Full-duplex operation* (sometimes just called *duplex*)—This describes a two-way system that can operate in both directions at the same time. The telephone system is full duplex, in that you can interrupt

the person talking using the second direction path.

Fig 12-6 provides examples of the three possibilities for a simple telegraph system using wires to interconnect the telegraph key(s) and the buzzer(s). In the half-duplex configuration, in order to receive a signal from the far end, you must close the switch across your end's telegraph key. If both stations have their switch open, no one hears anything, nor do they if both are closed. Analogously, if two taxi drivers both push their microphone buttons at the same time, neither will hear the other.

How Long Should it Take to Switch?

It won't be a surprise to find that different applications have different requirements for switching time. A radar system may be the most demanding. A typical radar system uses the same antenna for transmission and reception, and it must switch from transmitting a pulse to

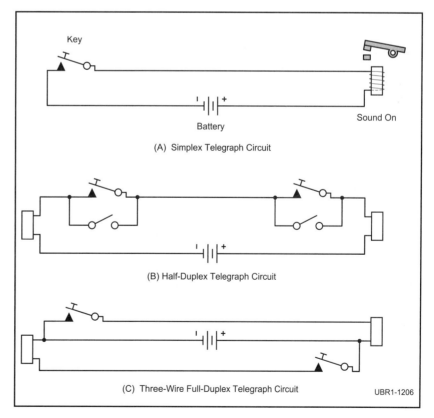

(A) Simplex Telegraph Circuit

(B) Half-Duplex Telegraph Circuit

(C) Three-Wire Full-Duplex Telegraph Circuit

UBR1-1206

Fig 12-6—Transmit-receive control. Three wire-line telegraph examples.

receiving a reflected pulse quite quickly. If it doesn't switch quickly enough it will miss reflections from targets close to the antenna. The speed of light is 186,000 miles per second, or 0.186 miles per microsecond. In that microsecond, the wave will go travel 0.093 miles out and back to the antenna. A radar engineer would round that to "10 μsec per radar mile."

Thus for every 10 μsec of lost switching time, we lose about one mile of coverage starting at the antenna. Since the pulse width may be 10 μsec long, we can't do much about the first mile, but we don't want to give up too many, especially if we are worried about aircraft landing at our airport! On the other hand, if we "jump the gun" and switch before the transmit pulse is completely over, we risk dumping lots of transmitter power into our sensitive receiver. This is not a good plan, especially if your radar runs megawatts of output power!

Other systems generally have less demanding requirements. A push-to-talk radiotelephone system probably will not lose much if it takes a few tenths of a second to switch over. Some two-way voice systems, especially those that don't transmit a carrier, such as SSB, use *voice-operated-transmit* or VOX operation. This allows operation that is close to full duplex, if both parties are polite. When the operator at one end stops talking, between words if not syllables, his radio automatically switches to receive, in case the operator at the other end wishes to make a comment. Unlike real full duplex, this doesn't work if both speak at the same time.

There are analogous systems for radiotelegraph, in two flavors. The *Tuna Tin 2* uses a straightforward half-duplex system. When one operator is finished with her transmission, she sends the letter "K," the telegraph equivalent of "over" and throws her switch from transmit to receive. The other station then switches to transmit and starts sending.

It is also possible to have an automated switchover in which a telegraph transmitter switches to transmit as soon as the key is pressed. If it switches back to receive as soon as the key is released, so the operator can hear between each dot and dash, this is called *break-in* keying. This can approach full-duplex operation. But just as in the case of VOX operation, if both send at the same time, neither can hear.

The other flavor is called *semi break-in*, usually using the same circuitry used for VOX operation. The switch to transmit occurs when the key is depressed. However, the switch to return to receive is delayed until after a pause, so no reception occurs between dots and dashes. This strategy is much like half-duplex, but eliminates the need for the operator to manually throw a switch to transition between transmit and receive.

Full-duplex operation usually requires a separation between the transmitter and the receiver frequency, plus extensive filtering to keep the transmitter's power from overloading the receiver. Alternately, geographic separation between the transmitter and receiver can be employed. In the days before high-capacity transatlantic cables, most telephone calls between North America and Europe were transmitted over HF. At the US end, the transmit and receive locations were hundreds of miles apart so that full-duplex operation could be used.

Notes

[1] In Chapter 17, I will describe how to best predict what range we can expect—not an exact science, especially at HF.

[2] The technology may well predate the light bulb, and the US Navy was not known for making changes easily.

[3] There is serious possibility for interference, especially if connected to an antenna. Causing such interference is in violation of FCC regulations with potentially serious consequences.

[4] This transmitter is based on the simple Colpitts LC oscillator shown in Figure 14.13(B) on page 14.14 of *The ARRL Handbook*, 2000 Edition.

[5] E. Hare, "The Tuna Tin 2 Today", *QST*, Mar 2002.

Review Questions

1. Describe how to change the transmit frequency of the simple transmitter in Fig 12-1. Repeat for the transmitter in Fig 12-3.

2. Why must the rise times of the circuits being keyed in a telegraph transmitter be carefully considered?

3. What might happen if the power supply voltage dropped every time a telegraph transmitter was keyed?

Chapter 13

Transmitting Voice

Some Amateur-Radio operators will travel great distances to be a "rare one." These adventuresome hams are at the top of Mount Kilimanjaro in Tanzania.

Contents

Amplitude-Modulated Voice

While the telegraph key in the transmitters of Chapter 12 can be considered a *modulator* of sorts, we usually reserve that term for a somewhat more sophisticated system that adds more information to the transmitted signal than just on-off Morse Code keying. As noted earlier, there are three signal parameters that can be used to modulate a radio signal and they all can be used in various ways to place voice (or other information) onto a transmitted signal. I will describe each modulation system in some detail in this chapter.

One possible way to add voice to a radio signal is to first convert the analog voice signal to digital data and then transmit it as "ones" and "zeros" of digital data. This is a technique employed for some applications using data applied to a pulse transmitter. However, here I will talk about the more straightforward application of analog voice to the amplitude of a carrier signal.

The form of voice amplitude modulation first employed in the early days of radio was called *high-level amplitude modulation*. This was generated by modulating an RF carrier with an audio signal. I described the conceptual view of this in Chapter 6. **Fig 13-1** is a repeat of the figure from that earlier chapter.

In **Fig 13-2** I have shown a more detailed view of how such a voice transmitter would actually be implemented. The upper portion is the RF channel. For example, you could think of the simple *Tuna Tin 2* transmitter from Chapter 12 as a transmitter, assuming that the crystal setting the CW frequency is changed to the voice portion of the band.

The lower portion is the audio or AF (audio frequency) channel, usually called the *modulator*. This is nothing more than an audio amplifier fed by a microphone. The output

power from the modulator is applied in series with the dc supply of the output stage of the RF channel of the transmitter. The level of the voice peaks must be just enough to vary

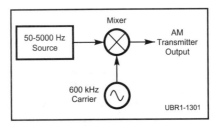

Fig 13-1—Block diagram of a conceptual AM radio transmitter.

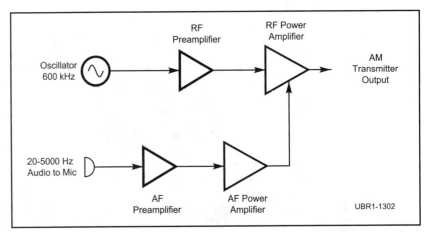

Fig 13-2—Block diagram of an actual 600-kHz high-level modulated AM radio transmitter.

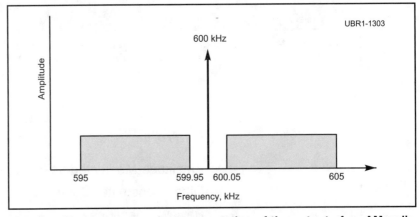

Fig 13-3—Frequency-domain representation of the output of an AM radio transmitter.

the supply to the RF amplifier collector from zero volts on negative peaks, to twice the normal supply voltage on positive voice peaks. This usually requires an AF amplifier with about half the average power output as the dc input power (the product of the dc collector voltage times the dc collector current) of the final RF amplifier stage.

The output signal occupies a frequency spectrum shown in **Fig 13-3**. The spectrum shown is that of a standard broadcast station, with an audio passband from 50 Hz to 5 kHz. Note that the resulting channel width is just twice the highest audio frequency transmitted. If the audio bandwidth were limited to typical "telephone quality speech" of 300 to 3300 Hz, the resulting bandwidth would be reduced to 6.6 kHz.

Note also that a perfect multiplication process (as described by Eq 6-1 at the beginning of Chapter 6) would create just the two sidebands and no carrier. The high-level modulation scheme in Fig 13-2, however, provides a carrier and the two sidebands from the product terms.

More Efficient AM Voice—Single-Sideband, Suppressed-Carrier Transmission

As we discussed in Chapter 9, the two sidebands of a standard AM transmitter carry (reversed) copies of the same information, and the carrier carries essentially no information. We can more efficiently transmit the information alone by transmitting just a single sideband and no carrier. In so doing, we use somewhat less than half the bandwidth, a scarce resource, and also consume much less transmitter power by not transmitting the carrier.

The block diagram of a simple single-sideband, suppressed-carrier

(abbreviated as *SSB*) transmitter is shown in **Fig 13-4**. This transmitter uses one of the balanced mixers from Chapter 6 as a *balanced modulator* to generate a double-sideband signal without a carrier. That signal is then sent through a filter designed to pass just one of the sidebands. The resulting SSB signal is next amplified to the desired power level, and we have an SSB transmitter.

While a transmitter of the type in Fig 13-4, with all processing done at the desired transmit frequency, will work, the configuration is not often used. Instead, the carrier oscillator and sideband filter are often at an intermediate frequency (IF) that is then heterodyned (mixed) to the operating frequency, as shown in **Fig 13-5**. The reason is that the

sideband filter is a complex narrow band filter and most manufacturers would rather not have to make a new filter design every time a transmitter is ordered for a new frequency! Most SSB transmitters can operate on different channels as well, so this avoids having to provide a bunch of expensive single-sideband filters, at a cost of an additional mixer, oscillator and one filter.

Note that the block diagram of our SSB transmitter bears a striking resemblance to the diagram of a communications receiver discussed in Chapter 10, except that it is turned around end-for-end. The same kind of requirements for image rejection that were receiver design constraints discussed in Chapter 11 apply here as well for a transmitter.

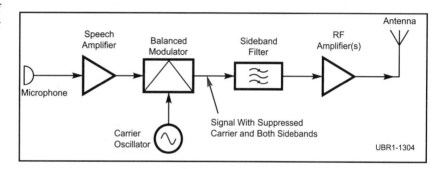

Fig 13-4—Block diagram of a filter type single-sideband, suppressed-carrier (SSB) transmitter.

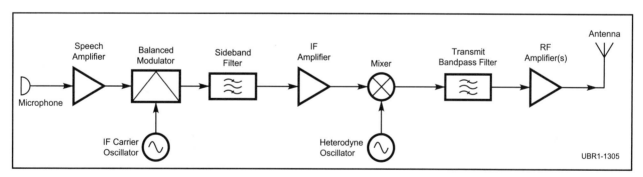

Fig 13-5—Block diagram of a filter-type SSB transmitter for multiple output frequency operation.

While many current radio communication transmitters are designed in the form of the SSB transmitter just discussed, there are other types that should be described, since you may encounter them in the field. I will briefly discuss the most significant in this section.

The Single Sideband Transceiver

Since the SSB transmitter and receiver are so similar, it is possible to combine them into a single package that shares many functions. A simple example is illustrated in **Fig 13-6**, in which common oscillators are used for transmitting and receiving. An advantage of this configuration in a two-way system is that the radios at both ends are set to transmit and receive on the same frequency as soon as an operator tunes his receiver to the frequency of the station at the far end.

In many transceivers, the sideband filter, some amplifiers and other filters are also shared between the transmitter and receiver. This is done through the use of extensive switching between transmit and receive.

While some fixed stations continue to use separate transmitters for maximum flexibility, the majority of HF two-way SSB installations, especially in the land mobile, aircraft and maritime services use transceivers. Two representative SSB transceivers from the amateur service are shown in **Fig 13-7** and **Fig 13-8**. Note that with appropriate switching of filters and other slight changes, most SSB transceivers are also capable of radiotelegraph, radioteletype and in some cases double-sideband (DSB) AM and FM voice operation.

SSB Phasing Method Transmitter

Most current SSB transmitters use the method shown in Fig 13-4 and discussed in the previous section to generate the SSB signal. That method is called the *filter method*, but really occurs in two parts. First, a balanced modulator is used to eliminate the carrier, and then a filter is used to eliminate the undesired sideband. The filter will often improve carrier rejection as well.

The *phasing method* of SSB generation is exactly the same configuration as the image-canceling mixer described in Chapter 6. This uses two balanced modulators and a phase shift network for both the audio and RF carrier signals to produce the upper sideband signal as shown in **Fig 13-9**. By a shift in the sign of either of the phase-shift networks, the opposite sideband can be generated. This method trades a few phase-shift networks and an extra balanced modulator for the sharp sideband filter of the filter method. While it looks deceptively simple, a limitation is in the construction of a phase-shift network that will have a constant 90° phase shift over the whole audio range. Errors in phase shift result in less than full carrier and sideband suppression. Nonetheless, there have been some successful examples offered over the years.

Independent Sideband or Multiplex Transmission

Some early SSB transmitters that were parts of systems designed for DSB AM were offered with independent sideband (ISB) capability. In this system, two audio channels are provided, each translated to one of the sidebands surrounding a (suppressed) carrier. Two audio channels were thus provided in the same channel bandwidth previously assigned to a single AM channel. This can be effective, but only for broadcast or full-duplex operation

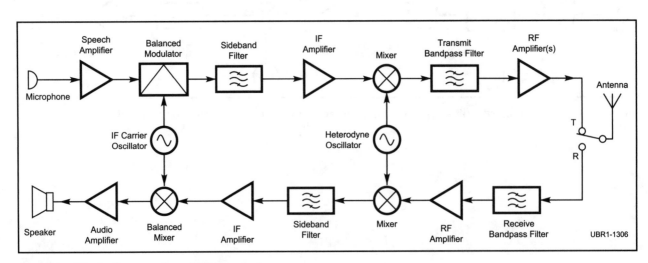

Fig 13-6—Block diagram of an SSB transceiver.

since it would be hard to imagine two two-way communications systems in which both users would be ready to switch from transmit to receive at exactly the same time.

The ISB idea quickly expanded into the more general *frequency-division multiplex* approach described in Chapter 7 and illustrated in Fig 7-9. The transmitter is designed to handle a bandwidth appropriate to the government-assigned allocated channel width, typically as an SSB transmitter using a baseband input (as described in Chapter 7). A multi-channel multiplex system, usually outside the transmitter proper, would supply a modulating signal of the appropriate bandwidth. The result is a transmitter carrying multiple voice channels suitable for broadcast or full-duplex operation.

Angle Modulated Transmitters

Transmitters using *frequency* or *phase modulation* are generally grouped into the category of *angle modulation* because the resulting signals are often indistinguishable, since an instantaneous change in either can appear identical, even though the method of modulating the signal is somewhat different. To generate an FM signal, you need an oscillator whose frequency can be changed by the modulating signal. Fortunately, there are only a finite number of different circuits around, and I've already discussed one that will do what you want. In Chapter 5, when discussing phased-locked-loop oscillators, I described an oscillator whose frequency could be changed by a "tuning voltage." Well, if you apply a voice signal to the TUNE connection point in Fig 5-3, you will change the frequency with the amplitude and frequency of the applied modulating signal, resulting in an FM signal.

The phase of a signal can be varied by changing the values of an R-C phase-shift network. One way to accomplish phase modulation is with an active element. See, for example **Fig 13-10**, in which the current

Fig 13-7—Photo of a modern basic SSB transceiver.

Fig 13-8—Photo of a modern advanced SSB transceiver.

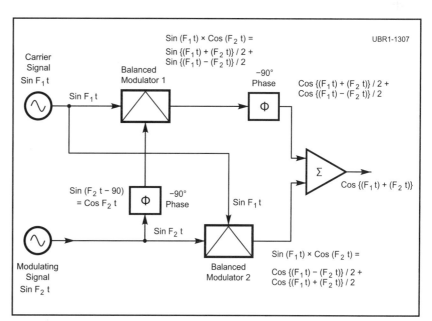

Fig 13-9—Phasing method of SSB generation.

through the field-effect transistor varies with the applied modulating signal to shift the phase and thus generate a PM signal. Because the effective stage load is changed, the carrier is also amplitude modulated and must be run through an FM limiter to remove amplitude variations.

Transmitters Using Frequency Multiplication

So far I have talked about changing transmitter frequency through heterodyning with different frequency oscillators to translate the frequency to another range. An alternative is to start with an oscillator frequency that is a fraction of the desired transmitter output and use *frequency multiplier* stages to get to the desired output frequency.

A classic frequency multiplier is just an RF amplifier stage with the input tuned to frequency f and the output tuned to $n \times f$, where n is an integer greater than one, typically two or three. This will only work well if the amplifier is designed to be non-linear, typically by biasing the

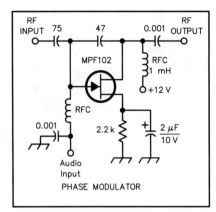

Fig 13-10—Phase modulator circuit.

stage beyond cut-off so that there is a large harmonic content. The output tuned circuit or filter is then adjusted to the desired harmonic and the desired result is obtained.

Note that this approach has a limited number of applications. In particular, it is not useable for AM in any of its forms since the amplitude variations do not make it through the nonlinear stage. If they did, each frequency component of the sidebands would be multiplied, resulting in a transmitted signal n times as wide as required by the information, and it would not be recoverable through normal receiver means.

The frequency multiplier is right at home, however, as a part of a radiotelegraph transmitter, since a carrier going on and off at frequency f sounds about the same as one at n × f, except for the frequency.

The major application for frequency multiplication these days is in FM transmitters, where the multiplication is used to advantage to obtain the desired frequency deviation. For example, a wideband high-fidelity FM broadcast transmitter seeking to fill a 150-kHz channel at a 100-MHz carrier frequency with music extending to 15 kHz would require a deviation of ±60 kHz [that is, the bandwidth BW = 2 × (M × D) as described in Chapter 10]. Such a transmitter might start with an 8.333-MHz oscillator modulated at an easy-to-generate ±5 kHz deviation. This is multiplied by 3, then 2 and then 2 again to end up at 100 MHz with ±60 kHz deviation.

Review Questions

1. Why is it important to receive the same sideband that is being sent by an SSB transmitter?

2. If the SSB transmitter of Fig 13-5 uses a IF carrier oscillator of 455 kHz, a sideband filter that passes from 455.3 kHz to 458.3 kHz, and a heterodyne oscillator of 10 MHz, what frequencies and sidebands come out of the mixer?

3. Describe two ways in which the opposite sidebands could be selected in such a system as in Question 2.

Chapter 14

Power Amplifiers, a Step Up

Inside view of Emtron DX-1d RF section. The quality of construction is evident.

Contents

In the last chapter, I discussed the various flavors of transmitters and transceivers that a radio worker might encounter. The designs had in common that they worked at a relatively low power level, with additional power provided by *power amplifiers* as might be needed. We have seen power amplifiers before. Even the two-stage "Tuna-Tin 2" transmitter from Fig 12-4 consisted of a crystal oscillator followed by a small power amplifier.

What is it that makes an amplifier into a power amplifier? Not much really, since almost all amplifiers produce higher power at their output than at their input, and thus could be correctly called power amplifiers. A comparison of the Tuna Tin 2 power amplifier shown in Fig 12-4 with the receiver RF amplifier shown in Fig 3-6 shows little difference. Since both are designed to increase the level of RF signals, that should not be too surprising! When amplifiers start being asked to deliver significant power to drive loudspeakers, servomotors or antennas, we start to call them power amplifiers. The distinction is rather arbitrary; however, the differences start to become evident as the power level gets higher and issues such as efficiency, thermal management, protection systems, distortion and filtering become significant.

Efficiency—What's the Scoop?

Most small-signal amplifiers are operated in what is called *Class A*. This means that the device—it could be a transistor or vacuum tube here—is operated so that the expected range of input signal level never exceeds the bias voltage, and where the output current is not driven so hard that the device stops conducting current at its output. If an input signal did push the device into non-conduction, there would be

clipping and distortion, generally not a desirable attribute for a small-signal amplifier. **Fig 14-1** shows the input and output waveforms of a capacitor-coupled Class-A amplifier. In this example, the amplifier bias network would have to cause a rest or quiescent collector current, of at least 40 mA to insure that the amplifier will stay in the Class-A operating region; that is, to make sure that the collector current is never cut off.

So why would we want to do anything different? Good question—I'm glad you asked! The implication of having the device biased at Class A is that there is current flow during all parts of the input cycle. In addition there is current flow, even during periods when the input signal

is off, such as when a telegraph key is up, or between syllables during an SSB transmission. During those periods, the amplifying device is dissipating power, even though no RF power is being generated.

The collector—or for a vacuum tube—plate, efficiency of an amplifier can be defined as the ratio of RF output power to the dc collector, or plate, input power. This can be found by multiplying the collector voltage times the average collector current. Note that this is not quite equal to the total efficiency of the amplifier, since other power consumers in the circuit are not considered; however, this is the generally accepted definition.

For a Class-A amplifier, it can be shown that the maximum efficiency is 33%. This is not a major concern

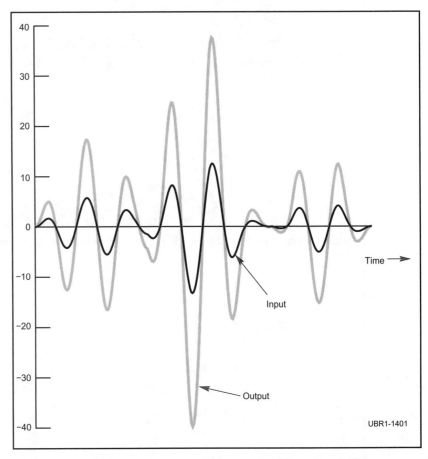

Fig 14-1—The input and output waveforms of a Class-A amplifier.

for small signal amplifiers, or even small power amplifiers. For example, if the $^1/_3$-W output Tuna Tin 2 amplifier stage were run in Class A, that would mean that the total collector dc power would be at least 1 W. A 12 V dc supply would have to deliver at least 83 mA to provide the output. The more efficient amplifiers that we will discuss next could deliver the same output with less than half the supply current. It doesn't sound like a lot, but for a portable transmitter, it would mean that the batteries for this stage would last twice as long, sometimes an important consideration.

The difference becomes even more dramatic as we get to higher power levels. Consider an RF power amplifier that delivers 1000 W. If the efficiency were 33%, we must supply 3000 W of dc power to the unit. This raises two concerns. First, if the amplifier runs on 12 V dc, a power supply must deliver a very hefty 250 A (a typical automobile storage battery might last ten minutes) to provide the needed power. Second, if we put in 3000 W and get 1000 W out, there's 2000 W left to be dissipated as heat. Extreme measures must be taken to remove the heat. Imagine the heat from 20 fully illuminated 100 W light bulbs in a small box!

Now, if we could achieve 75% efficiency, only 1,333 W would be needed from the power supply. This would require *only* about 111 A, instead of the 250 A for the previous case. Note that rather than having to dissipate 2000 W of heat we must eliminate 333 W. That's still a lot, but a reduction of 83%.

So How do we do That?

It probably won't come as a surprise that by adjusting the amplifier's bias we can change the conditions and class of amplifier operation to improve the efficiency. Let's first examine another topic.

Linear versus nonlinear amplifiers

The previous discussion about output power versus efficiency does raise some other issues. A Class-A amplifier is inherently *linear*. That is, the output signal is a faithful representation of the input, only more powerful. Recall, please, our earlier discussion of nonlinear devices, where a linear amplifier is one with a transfer function of $Y = K \times X$. Here,

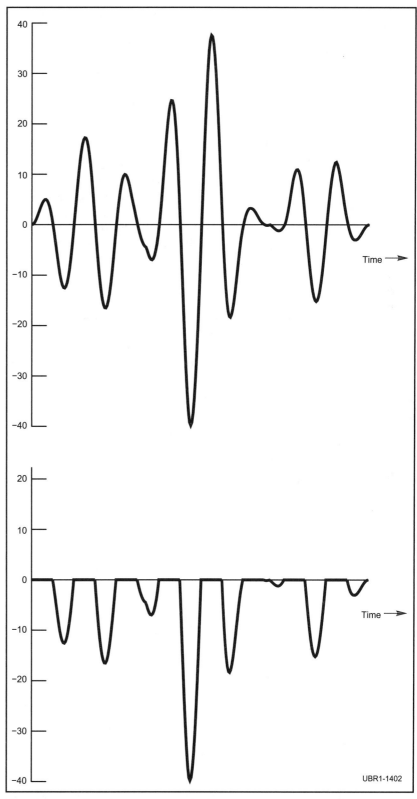

Fig 14-2—The input and output waveforms of a Class-B amplifier.

X is the input; Y is the output and K is the *gain* of the amplifier. The transfer function from input to output is a straight line, with a slope K. Hence, we use the term linear to characterize this system.

High efficiency amplifiers tend to be nonlinear. They put out lots of power on the frequency of their input signal whenever the input signal is present at sufficient amplitude. If the input is below a threshold established by the bias level they put out nothing.

To amplify signals that convey amplitude information, such as AM or SSB, we must use *linear amplifiers*. If we want to increase the power level of signals that don't contain such amplitude information (such as FM or on-off keyed AM, radiotelegraph or data), we can use higher efficiency, but inherently non-linear, amplifiers.

Higher Efficiency Linear Amplifiers

Fortunately, we aren't restricted to Class A if we want an amplifier to be linear. The next level in efficiency is called *Class B*. A Class-B amplifier has its bias set to just below conduction when no signal is present. For a common-emitter transistor amplifier there is no base-to-emitter current—and hence no collector current—when no input signal is present. As soon as a signal is applied, current begins to flow and the output follows the level of the input, only larger—just what we want, almost! **Fig 14-2** shows the input and output waveforms of a Class-B amplifier.

The good news is that the Class-B amplifier can be up to about 66% efficient. The bad news is that we have a new problem to deal with. While the negative values are a larger

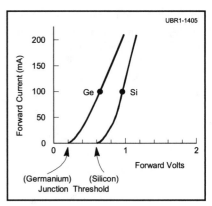

Fig 14-5—PN junction current as a function of applied voltage.

copy of the input, the positive values are gone. This is because a negative input current cannot cause the zero-biased base-to-emitter junction to conduct, and there is no collector current as a result. If this were an audio power amplifier, it would mean that the loudspeaker would just move outwards, never in the other direction. Audio would be a distorted, and would not look very linear.

It can be shown that any complex waveform can be represented as a series of sinusoids, with a fundamental frequency and various amplitudes of harmonics, just as we discussed in our earlier development of intermodulation distortion. The effect is called *harmonic distortion*. If this were an RF amplifier, a connected antenna would radiate not just the desired signal but signals at multiples of the desired frequency at different amplitudes. This would not be a good plan!

Taming the Class-B Linear Amplifier

One way to can get rid of harmonics is to use a filter to eliminate the undesired frequencies. As hard as it may be to imagine, if the second and higher harmonics of the signal shown in Fig 14-2 are eliminated with a filter, the resulting signal will look a lot like that of Fig 10-1. There is, however, an even more clever way to solve this problem—the push-pull amplifier.

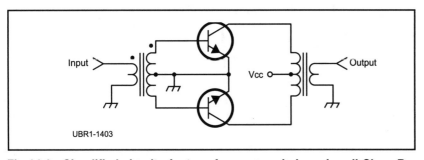

Fig 14-3—Simplified circuit of a transformer-coupled, push-pull Class-B amplifier.

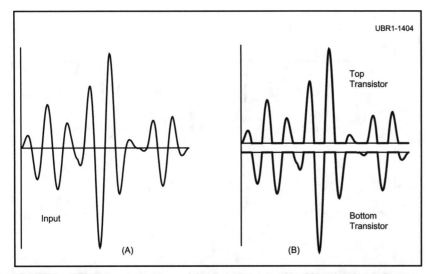

Fig 14-4—Collector current waveforms of the top and bottom transistors of an ideal push-pull Class-B amplifier.

Introducing the Push-Pull Amplifier

A simplified circuit of a transformer coupled Class-B amplifier is shown in **Fig 14-3**. It is simplified because it assumes ideal transistors whose collector-to-emitter junctions will conduct when any current flows in the base-emitter junctions. No bias circuit is shown in this figure; this will be discussed later. Note that this circuit could be either an audio or an RF power amplifier, depending on the frequency response of the transformers.

During times when the input signal is positive, the transformer will deliver positive current to the top transistor and negative current to the bottom transistor. During that interval the top transistor will follow the input, with the collector current an amplified version of the input signal, as shown in **Fig 14-4**. When the input signal is negative, the bottom transistor will receive a positive input and will conduct. The two signals will combine in the output transformer to produce a signal like that in Fig 14-1, except the efficiency will be higher, since there is no collector current when there is no signal present.

Looks Pretty Good—What Makes it Less Than Ideal?

Well, we're stuck with using real-world devices, and they have limitations set by device physics. A look at **Fig 14-5** will show the real conduction characteristics of PN junctions such as the base-emitter junction of the transistors in our less-than-ideal amplifier. Most power amplifiers use

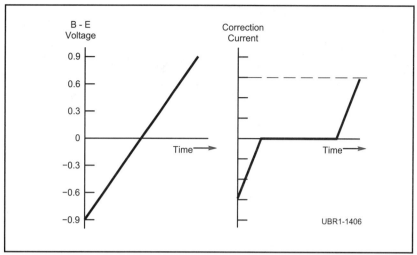

Fig 14-6—Illustration of crossover distortion. At A, input voltage, at B, output current.

silicon devices, so we are stuck with a region on each side of the input waveform of the transistors during which no current will flow. This region extends to an input voltage of ±0.6 V, resulting in a new kind of distortion as the signal crosses "no man's land" around the zero voltage level. This is referred to as *crossover distortion*. The effect is shown in **Fig 14-6**, the input voltage on the left, the output current on the right, with the scale exaggerated for effect.

The effect can be significantly reduced by adding a bias circuit that causes each transistor to be forward biased to just to the point at which it is about to conduct when no signal is present. While this helps, a look at Fig 14-5 will indicate that the junction response is not quite a straight line near the point where conduction starts, and thus while crossover distortion can be reduced

in this way, it can not be entirely eliminated. This means that to meet US federal regulations for harmonic output, some filtering will still be required.

If we continue to increase the bias current, so the transistors move up to the more linear portion of their operating curve, we tend to make the amplifier more linear. The downside is that we make the amplifier somewhat less efficient as well. The results are somewhere between the efficiency of Class-A amplifiers and Class-B amplifiers. By no coincidence at all, such amplifiers are called Class-AB amplifiers! There are even definitions of sub-classes in Class AB operation. Class AB_1 is closer to A, while Class AB_2 is closer to B. There is a specific difference defined in the next module—for vacuum tube amplifiers.

OK, we've put it off as long as we could. It's time to face the horrible truth—there are still places where vacuum tubes are king! Arguably, anything that can be done with a solid-state device can, and used to be, done with a vacuum tube. Starting with the invention of the transistor by three Bell Labs' scientists in 1947, there has been a migration away from tubes and towards transistors and even more exotic solid-state devices.

This is as it should be. Writing as a person whose career and experience spans both generations, most of the time I find it great to work in the solid-state environment. Solid-state devices are more reliable—tubes have filaments like light bulbs, and need to be replaced almost as often; tubes break when you drop them; tubes get hot even when nothing else is going on and tubes use dangerous high voltages that can make you jump, or worse, if you put your fingers in the wrong places!

On the other hand, tubes can handle more power than most solid-state devices. Here we are in the Power Amplifier chapter, so we better talk about tubes. Tubes do have some other advantages as well. They tend to be more forgiving—briefly exceed their ratings and they may glow more brightly than usual for a while, but they still work—not so transistors. They also are less subject to transients such as those caused by the effects of lightning or nuclear electromagnetic pulse (NEMP—glad I don't have to worry about that depressing cold-war issue any more!).

With each generation, transistors are able to move further and further into the vacuum-tube domain. Check your local listings, but as this is written, solid-state power amplifiers are easily putting out power in the single digits of kilowatts range—

although often using more than one device sharing the load within the amplifier. In some cases, input power is divided among multiple lower-powered amplifier modules and then recombined at the output. While high-power capable tubes are not inexpensive, high-power, RF-capable solid-state devices tend to be even more pricey. Above a few hundred watts, it is common to see amplifiers of either technology. The lead photo is of an HF linear power amplifier with an output rating of 1000 W peak envelope power (PEP) from 1.8 to 30 MHz. It requires 50 W of input power for full output and does it with a single vacuum tube.

So What are These Tube Critters Like?

The tube is the round object in the upper left corner of the picture at the beginning of this chapter. This tube is of ceramic rather than glass construction. The anode, from which any undelivered power must be dissipated, surrounds the outer surface and is attached to a honeycomb of vertical fins. Forced air is blown up past the fins to dissipate the heat. Other tubes, especially those at the tens of kilowatt level and above, may have a jacket around the anode to allow for pumped water cooling with

an external heat exchanger, very similar to the radiator on a car.

A simplified vacuum tube power amplifier circuit is shown in **Fig 14-7**. Note that unlike generally low impedance solid-state amplifiers that use wide-band transformers at the input and output, many vacuum tube circuits use high-impedance resonant L-C coupling circuits instead. In general, they need to be carefully tuned to the operating frequency, but they also remove much of the harmonic energy at the same time that they pass on signals at the desired frequency.

Some typical values may be of interest, **Table 14-1** shows representative conditions for a tetrode amplifier using two tubes

Table 14-1

Pair of 4CX1000A tube in Class AB$_1$ operation

Plate voltage	2500 V
Maximum plate current	2 A
Minimum plate current	0.05 A
Screen grid voltage	325 V
Maximum screen current	0.06 A
Control grid bias	−55 V
Filament voltage	6 V
Filament current	25 A (both tubes)

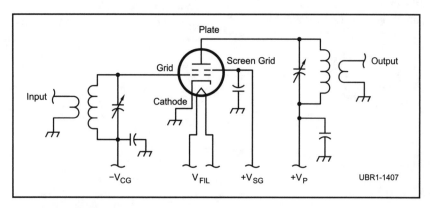

Fig 14-7—Simplified schematic of a tetrode vacuum tube power amplifier.

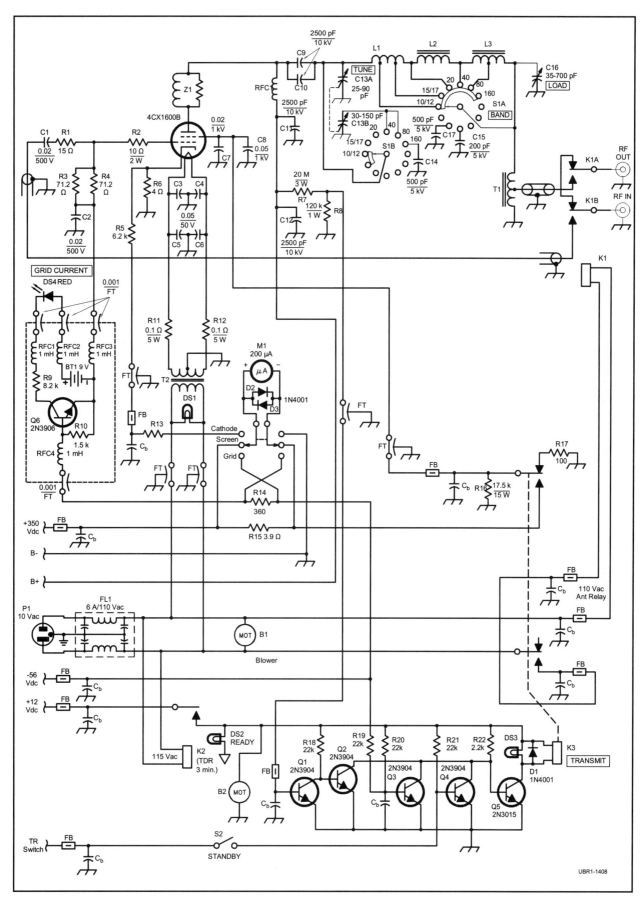

Fig 14-8—Schematic of a real tetrode vacuum tube power amplifier.

(4CX1000A) at the 3000-W output level: The –55 V grid bias establishes the operation at Class AB$_1$.

Some comments may be in order for those not too familiar with vacuum tubes and how they operate. First, it's very important to exercise all safety considerations when working around equipment like this. With a solid-state power amplifier, the 12 or perhaps 50 V dc generally encountered, can be somewhat dangerous (your author has had watchbands and wedding rings melted from contact with such supplies with high current capability). With a vacuum tube amplifier with 2500 V dc running around, you usually die from your first encounter. Please make sure power is not just off, but disconnected and that all capacitors are discharged using an insulated and grounded shorting stick before venturing inside.

Some less dire comments: The filament is about the same as one in a light bulb, hopefully longer lasting, and only serves to heat the cathode. The 150 W of heat generated by the filament must be dissipated along with all other heat sources. The plate is the element that acts very much like the collector in a transistor amplifier. The plate current is controlled by changes in the control grid voltage. Think of the screen grid as an accelerator. While the dc bias on it has an effect on the properties of the tube, the bias is generally static once its value is set.

A vacuum tube Class-AB$_1$ amplifier is defined as an amplifier between Class A and Class B in which the grids never get driven to a positive voltage. On the other hand, a Class AB$_2$ tube does get driven into grid current by the peaks of the input signal. Based on the table above, we can see that at no signal, (grids at –55 V) the amplifier's plate dc current is 0.05 A. At peak input, the grid is driven to 0.0 V, and the plate current goes to 2.0 A.

How About the Real Deal?

Fig 14-8 is the schematic[1] of a 1500 W PEP output HF power amplifier designed for home construction by advanced amateurs. We won't go through all the design details here, but will cover the high points. Note first that the actual amplifier covers just the quarter of the page. The rest of the schematic deals with support circuitry, including that for safety (of the tube, not the people!) so that the amplifier will shut down if key limits are exceeded. There is also metering so that conditions can be observed during operation. There is control circuitry, transmit-receive switching and amplifier bypass when it not needed. Note that the power supply required to provide all the required voltages fills up another whole schematic page.

This amplifier is somewhat similar to the one in the lead photo, but does have some design features that contrast it from our simplified circuit, as well as from the one in the photo. The output-tuned circuit consists of a combination equivalent L-C resonant circuit (a Pi configuration, in this case), combined with a wide-band transformer. The transformation from high output impedance to the 50-W load in this amplifier takes place in two steps. The output circuit looks complicated because it is switchable to cover multiple frequency ranges. Another difference in the output circuit is that the dc is shunt-fed to the plate instead of connecting through the L-C circuit. This doesn't change operation except to keep the high voltage dc from appearing on the output circuit components. Another difference is in the input circuit. This tube has enough gain that by loading the input with a resistive network, the required input voltage can be obtained so an L-C circuit is not needed.

Can we get even less linear?

You bet! For many services linearity is less important than efficiency, and power amplifiers can be made much less linear than we have discussed so far. As I mentioned previously, radiotelegraph signals have no need to be amplified linearly, as one example. If the key is down, we want a burst of power output and if the key is up we want none. An amplifier that is not linear at all can support this input/output relationship.

The class-C amplifier

What comes next? It almost has to be a *Class-C amplifier*! The Class-C amplifier is biased well beyond cut-off (for a vacuum-tube amplifier), or heavily into the non-conducting region for a solid-state device. If we think about the part of a sinusoid that results in collector or plate current flow, a Class-A amplifier has current flow for 360° of the time; a Class-B amplifier for 180° of the time, and an AB-amplifier somewhere in-between. On the other hand, a Class-C amplifier conducts through 90° (or less) of the full cycle. Current in the output circuit flows only during a quarter or less of each cycle of the input waveform. **Fig 14-9** shows the input and output waveforms of a

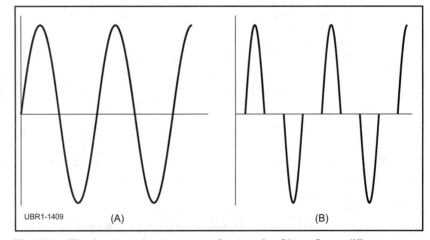

UBR1-1409   (A)   (B)

Fig 14-9—The input and output waveforms of a Class-C amplifier.

push-pull Class-C amplifier.

A Class-C amplifier can have an efficiency of up to about 75%, an improvement over our linear brethren. It cannot be used in applications requiring linear service, however, such as SSB or other "variable amplitude" modes. It can be used for FM or pulse applications—including radiotelegraph, data and radar. It also can be used as the final RF stage of an AM transmitter using a high-level modulation scheme, such as was shown in Chapter 13.

Note that my statements about classes of amplifiers are not device-dependent; each class can be implemented either in solid-state or vacuum-tube technology. One limit on the previous statement is that the distinction between classes AB1 and AB2, based on whether or not "grid current" flows, apply only to amplifiers with grids—that is, to vacuum tubes!

Also note that I haven't had to differentiate between circuits when I talk about amplifier classes. In general, the same circuits can be made to operate in any of the classes I've discussed so far—only the bias voltage (vacuum tube) or bias current (transistor) needs to be changed to change the class of operation.

I should also mention Field-Effect Transistors (FET). FETs are solid-state devices that act very much like vacuum tubes, because they are controlled by a gate that acts much like the grid of a vacuum tube. Some FETs are used as power amplifier elements and they share some characteristics of both tubes and transistors.

We Got Some Letters Left!

OK, you broke the code: after Class-C amplifiers come Class-D amplifier, and even some others beyond that. Beyond Class C, we get into what is called *switching-mode amplifiers*. These amplifiers start acting a bit like digital switches—when the input exceeds a certain threshold, the collector or plate current rises instantly to full value. It stays there through each cycle of the input waveform until the input signal drops below the threshold and then the collector or plate current drops to zero. These act like Class-C+ amplifiers on steroids and can be very efficient with up to 90% or even more of the dc input power leaving as RF output.

You won't see many amplifiers above Class C in typical radio applications, but they will appear from time to time in very high-power applications including radar.

Notes

[1]*The ARRL Handbook for Radio Communications*, 2005 Edition, Chapter 18, pp 29-35. Available from the ARRL Bookstore. Order number 9280. Telephone toll-free in the US 888-277-5289, or 860-594-0355; **www.arrl.org/shop/ pubsales@arrl.org**.

Review Questions

1. Under what condition can an amplifier with a voltage gain of 1.0 be a power amplifier? Consider an amplifier with a power input of 1.0 W, a power output of 10 W into 50 Ω and a voltage gain of 1.0. What is the output voltage? Input voltage? Input impedance?

2. Consider the values in Table 14-1 above describing 4CX1000A tubes in AB$_1$ operation. If these were run at full input (and output) for a period of time, what would the plate input power be? If the output were 3000 W, what would the plate efficiency be?

3. Repeat the above calculation considering all the applied voltages and currents for each grid. What would the amplifier (rather than plate) efficiency be?

Chapter 15

Propagation—Getting Your Signal Where You Want It

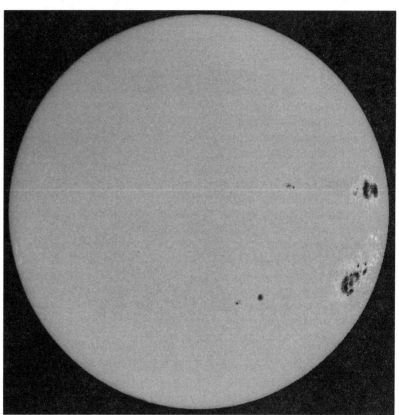

When there are sunspots on the sun, the ionization level in the Earth's ionosphere is boosted—providing better communications on the higher HF bands.

Contents

One of the facets of radio that make it an interesting topic is the manner by which signals get from one place to another. Many signals can be described by the *line-of-sight* path that works just like light. It starts here and goes there, and you can see one end from the other. Just as in the case with light, the brightness of the light at the end depends pretty much on how bright the light is when it starts, how far apart the ends are and how sensitive the eye is at the far end.

A line-of-sight path is pretty easy to determine with just a bit of analysis. You must know the height of each endpoint and how much the earth curves in-between to determine the maximum distance. Note that if there are obstructions in the path, it is no longer line-of-sight and we have to look toward other means to get the signal through. As with light, radio signals can reflect off objects (especially metal, but also water, for example) and refract (bend) while traveling through some objects, just as light does going through water.

The intensity of the light—or a radio signal—as it moves from the source to the observer appears to diminish as the distance becomes larger. This is true even if the signal is not being absorbed by anything in between, but is simply due to the amount of energy available to your eye—or to the receiving antenna—as the energy spreads out over a wider area as it goes. Visualize (in your mind or actually try an experiment) a candle or a flashlight bulb putting out light in all directions. Let's assume that it is a 1-W bulb. How can we determine how much energy will be there for a receiving area at a certain distance from the source? If the light goes out in all directions, we can examine the total light as if passing through an imaginary sphere at any distance from the source. The surface area of a sphere can be determined if we know the distance from the source. From geometry we know that the area of a sphere is $4\pi r^2$, where r is the radius.

Let's find out how much light intensity is available at two distances. First, at 10 meters away the area of the imaginary sphere is 1256 square meters. The 1 W of light is spread evenly over the whole surface, so the power available in a square meter is 1/1256 or about 0.0008 W/m², which can also be expressed as 0.8 mW/m². If your eye picks up light from a square centimeter of surface area, it will receive 0.8 mW/10,000 cm²/m², which is 0.08 µW in each square centimeter

around the sphere.

If you double the distance to 20 meters, you can perform the same calculations. The *power density* (the power per unit area) is now 2 mW/m² or 0.02 µW in your eye. Notice that because the area is proportional to the square of the distance, the power density also falls off as the square of the distance. If you knew the minimum amount of light you could see, you could easily calculate the maximum line-of-sight-distance at which you could still see the source, under the assumption that there was nothing in the way to block the light. See **Fig 15-1**.

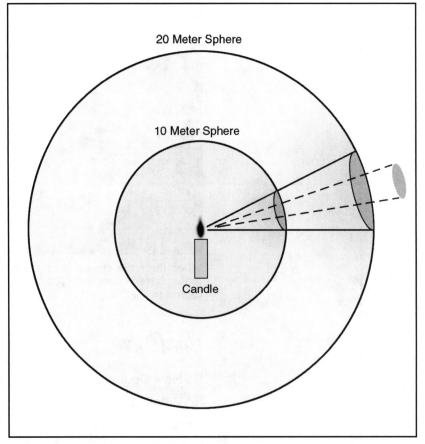

Fig 15-1—Illustration of reduction of signal energy with distance from source.

Suppose we place a mirror behind the bulb—how would our results change? If the mirror is a good one, and is properly placed, the energy that would have gone in the direction of the mirror would now reflect away from the mirror. In a perfect world, all the signal radiated in the direction of the mirror would be reflected, meaning that there would be twice the power in the direction away from the mirror reflected back toward the source. Thus a receiver 10 meters away behind the mirror would encounter an energy density of zero mW/m². The receiver on the side in front of the mirror would see a power density of 1.6 mW/m². An observer on the this side could easily assume that someone had "turned up the light" and couldn't tell the difference between a 1-W source with a mirror and a 2-W source without one.

Note that the power density is reduced in the same way as you move away from the reflector. If you look at the lens in a flashlight, you will likely see a more complicated reflector than just a flat mirror. The typical reflector is a *parabola*, which takes the light from the *focal point* (where the bulb is positioned) and then focuses the light into a much narrower beam. The light appears brighter to an observer in the beam (or an object in the beam) than would be the case with a flat mirror.

Signals that leave the transmitting antenna in an upward direction are called *sky wave signals*. They still travel in a line-of-sight fashion, such as in ground-to-air communications, but the most interesting ones to me are the HF signals that interact with something called the *ionosphere*. The ionosphere is a region surrounding the earth at a height of roughly 30 to 260 miles (see **Fig 15-2**). Because this is farther away from the pull of gravity than our atmosphere, the molecules encountered by radio signals at ionospheric heights are less dense than they are down where we can breath. The density of air molecules encountered by the radio signals is reduced as the signals move from the bottom of the ionosphere towards the upper reaches, until there are almost no molecules left, just as in deep space.

It is convenient to talk about the ionosphere in terms of *regions* or *layers*, since the effect on radio waves varies with the height of the ionosphere. In all portions of the ionosphere energy from the sun during daylight periods interacts with the molecules as shown in **Fig 15-3**. Ultraviolet radiation has the biggest effect; however, X-rays and other forms of radiation interact with the air molecules as well. The effect of radiation is to raise the energy in the molecules, causing them to release electrons. When an electron is released from an electrically neutral molecule (that is, one that has the same number of positive and negative particles) we end up with an electron (negative charge) and a positively charged *ion*. A radio signal encountering an ionized area will react in one of the following ways, depending on its wavelength and the density of ions.

1. It can be absorbed. This will occur if the ion density is high and the wavelength is long.

2. It can be transmitted through the ionosphere without much loss. This will happen if the wavelength is short compared to the space between ions. It also depends on the angle at which the wavefront strikes the ionized layer.

3. It can be bent, or *refracted*. This happens if the wavelength is between the other two cases. As the wavefront continues to be refracted, it tends to bend down away from the highest concentration of ions and returns to earth. Even though it is a refraction rather than a reflection, the effect is similar to a reflection from a surface slightly higher than the actual refracting area.

Ionization occurs because of the energy of the sun, and it happens only on the side of the Earth in daylight. As an area on the Earth moves from daylight into night, the electrons tend to recombine with the ions, returning air molecules to an electrically neutral state. All areas do not recombine instantly though. The recombination occurs in a rather random fashion with electrons recombining with ions they happen to run into. As a consequence, denser areas, closer to the Earth's surface, recombine more quickly, and those further away take longer, sometimes staying ionized throughout the night.

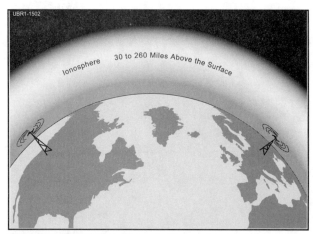

Fig 15-2—The ionosphere is the region above the Earth's surface (not to scale).

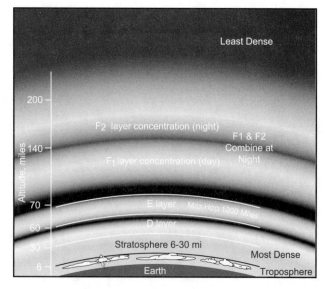

Fig 15-3—The layers of the ionosphere.

Layers of the Ionosphere

The ionosphere doesn't really have "layers." Spaceships don't observe bumps or roadsigns as they pass through, but people talk about *ionospheric layers* because of how the ionosphere interacts with radio waves passing through them differently at different heights. It's convenient to group the regions together as if they formed discrete layers. Each layer is labeled with a letter starting at our atmosphere and continuing up through the ionosphere. Because pioneers in the study of radio propagation started out with the letter E, (standing for "electric layer"), there are no A, B and C layers. The D, E and F layers make up the parts of the ionosphere that generally effect radio communications. These layers and their approximate distances are shown in Fig 15-3.

What About Antennas?

We mentioned earlier that radio signals act much like light. Just as light can be reflected from a mirror, radio waves can be reflected, but by an "electric mirror." The actions of such a radio-signal reflector can be very much like those that reflect light. For example, the reflector behind the bulb in a flashlight, or car headlight, has a shape similar to the reflector behind a radar antenna. The difference is, the radar reflector is made from a conducting material such as metal, while the light reflector is usually polished metal, or it could be a non-conductive material with a shiny surface.

If you spend a lot of time looking at radio reflectors, you will notice that they aren't always solid—sometimes they are made of screen-like material. If these were used as light reflectors, some of the light would leak through. Why doesn't this happen to radio waves? The reason is that light is at a very high frequency (a short wavelength) compared to radio frequencies. The designer of a radio reflector selected the size of the holes in the reflector so that they would be small in terms of the radio signal's wavelength and thus the radio signal doesn't see the holes at all. We can consider a reflected signal a sort of *bent* line-of-sight signal. We'll talk more about antennas later in this chapter and again towards the end of the book.

Something that happens with low-frequency radio waves that doesn't happen with light is called *ground-wave propagation*. Here the radio signal tends to be absorbed as it passes over the ground, resulting in the signal being tilted. Think of it being *dragged down* as it travels along. The dragged-down signal hugs the ground and continues along to provide coverage beyond the line-of-sight distance. The signals get weaker rather quickly, however. This is the way high-powered AM broadcast transmitters in the 530 to 1700-kHz range can reach radios 50 to 100 miles away, far beyond the line of sight.

Again, the reason we divide the ionosphere into layers is that each region tends to impact radio transmission differently. In addition, just as the size of holes in a reflector determine which frequencies get reflected and which pass through, the action of the ions in the ionosphere depend on the relationship between spacing of ions and radio wavelength. Thus, each ionospheric effect acts differently for different frequencies .

If those were the only changes we had to worry about, this would be merely difficult! However, we have one more major effect to consider here and that is changes in the sun. The radiation levels of UV and X-rays from the sun cause the ionization in the ionosphere. While we are pretty good at predicting when it will be day and night, we are less able to predict the level of non-visible energy leaving the sun. In addition to the sunlight that we enjoy most days, the sun has frequent hydrogen-bomb-sized eruptions from areas called *sunspots*. Sunspots have the appearance of black spots on the surface of the sun. (*Never* look directly at the sun or your eyes can be damaged. Instead, focus the image of the sun onto a piece of white paper.)

Sunspots are major contributors of energy that cause ionization of our ionosphere, some 93 million miles away from the sun. People have been observing and recording the presence of sunspots on the sun for hundreds of years, long before they worried about radio propagation. By looking at the historical data, we see that the average number of sunspots changes in an approximately 11-year cycle. We can't accurately predict the quantity of spots on any given day, but we can predict general trends. For this reason, the only way to be sure how the ionosphere will act at a particular time is to test it, either indirectly (by counting sunspots), or

directly (by sending signals into it). We can discern some general characteristics according to the ionospheric layer and the frequency range to give an idea what to expect.

How About Those Layers?

Ionization in the D layer recombines quickly at sunset because it is dense compared to the layers further from the effects of the earth's gravity. Thus, the D layer has impact only during daylight hours. HF and above signals tend to pass through the D layer, while MF and lower frequencies tend to be absorbed. It's the D layer that causes MF broadcast-band transmission to be limited to ground wave during the day rather than allowing sky-wave propagation.

HF signals that make it through the D layer can be impacted by the E layer. This can refract HF signals and result in medium distance (500 to 1000 mile) communications. At some times even VHF signals will refract via the D layer, but this is an unusual phenomenon.

The F layer, the highest above the

earth, is responsible for the longest-range communication, with hops of 2500 miles not unusual. Since the F layer can stay ionized even after dark, the propagation can be very good then, since the absorption of the D and E layers is minimal after they have recombined when the sun goes down. Longer ranges are possible through multiple-hop paths.

Since the particular frequencies at which these effects change on a daily and hourly basis, the frequencies are identified with particular names as follows:

• The *Lowest Usable Frequency* or LUF is the lowest frequency that can make it through the D layer and be refracted to reach a particular destination. This term is generally destination-specific, so some far-away destinations requiring long-distance hops by means of F layer propagation will have a higher LUF than other paths that could be supported by the E layer. Frequencies below the LUF tend to be absorbed, as shown in **Fig 15-4**. Frequencies above the LUF tend to

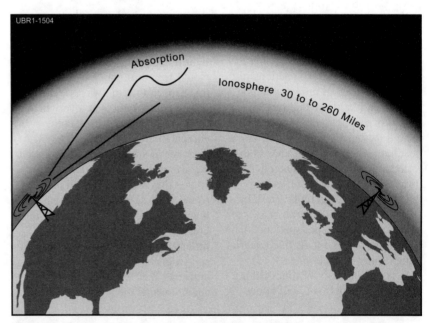

Fig 15-4—Propagation at frequencies below the LUF (lowest usable frequency).

be refracted or bent back to return to the earth long distances away. Down on the lower HF frequencies, received noise from lightning discharges and other such atmospheric noises limits the LUF.

• The *Maximum Usable Frequency* or MUF is the highest frequency that will be refracted on its way to a destination, rather than continuing on through the ionosphere to be lost in outer space. Frequencies between the MUF and the LUF will generally be useful for a specific path, although the higher the frequency in that range, the stronger the received signal will tend to be since there will be less absorption. See **Fig 15-5**.

• Communication in the region between the LUF and MUF generally occurs in the shortwave region of the HF spectrum. This provides a useful method of communication; however, it requires a certain level of skill to take advantage of it, since the useful frequencies change due to sunspot activity, the month of the year and during each day as well. This kind of propagation is referred to as *skip*, since the region between the transmitter and receiver includes a *skip zone* in which there is no reception. This is illustrated in **Fig 15-6**.

• In addition, there is a *critical frequency* or F_c and an associated *critical angle*, above which a signal will plow through the entire ionosphere, as shown in **Fig 15-7**. This angle is very much like the angle above which you can see the bottom of a pond and below which you see a reflection of the trees and sky above—again, light and radio waves share some characteristics. Both the critical frequency and the critical angle change with ionospheric conditions. Satellite and space communication require transmission above the critical frequency and critical angle. **Fig 15-8** shows some types of space communication.

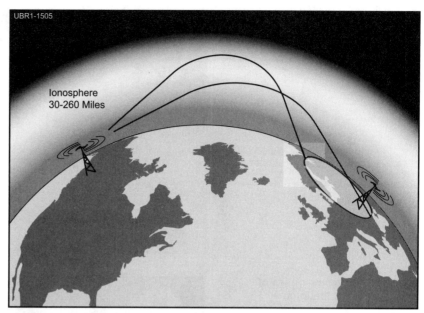

Fig 15-5—Propagation of signals between the LUF and MUF (maximum usable frequency).

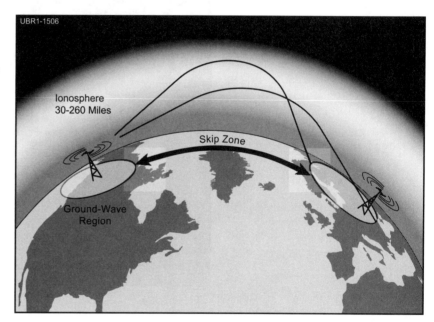

Fig 15-6—The region between the ground-wave coverage area and the area where refracted signals are present is known as the *skip zone*.

Fig 15-7—Propagation for frequencies above the critical frequency or critical angle.

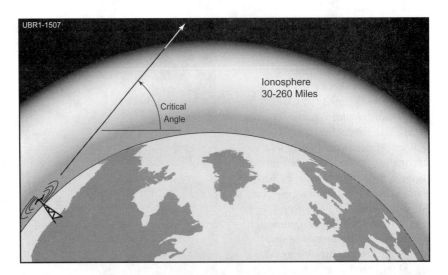

Fig 15-8—Some examples of space communications.

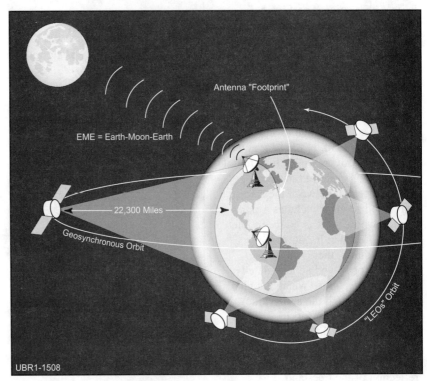

Review Questions

1. I mentioned that radio waves act a lot like light. In fact, before radio was around light was used as a mechanism for medium to long distance communications. Can you think of three types of early light-based communication systems?

2. Consider a world without an ionosphere. How would radio communication be different than in our world?

3. Describe why ionospheric communication can not always be counted on to behave in a certain way.

4. Why do long range HF communications systems usually have a number of frequency assignments in different parts of the HF spectrum?

Chapter 16

Transmitting Antennas

Transmitting antenna for quick setup and takedown in emergency portable operations.

Contents

Any piece of wire or tubing will work as an antenna if you could cause a time-varying current to flow through it. The trick is to get the current to flow and have the return current not cause an opposite electromagnetic field to cancel things out.

While there are many exceptions and special cases, this is easiest to do with a structure comparable in size to half a wavelength at the frequency you want to launch. In this discussion, it is important to not confuse the *antenna* with the *support structure*. For example, the antenna for a MF AM broadcast station is likely to be a vertical tower from 150 to 500 feet tall. In this case the tower *is* the antenna. A VHF FM broadcast station will have an antenna that is 10 to 20 feet in height. But to obtain a long line-of-sight path to its listeners, it may be located on top of a tower that is 500 feet high. Unless you look carefully it may be hard to tell the difference!

The Dipole Antenna

The simplest and one of the most frequently encountered types of antenna is called a *half-wave dipole* and is shown in **Fig 16-1**. It is merely a length of wire or rod approximately an electrical half-wavelength long, generally split in the center and connected to a transmitter sending signals on a frequency corresponding to the half wavelength. In many cases it is desirable to have the transmitter and antenna located in different places. In this case they are connected by a *transmission line*, which I will discuss in a subsequent chapter.

The length and the conditions at the ends (*boundaries*) of the antenna determine the current and voltage along the antenna wire. Because the ends are not connected to anything, the current there is at a minimum

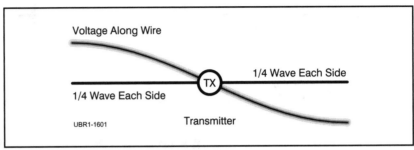

Fig 16-1—The dipole antenna. A "half-wave" antenna is easy to use, easy to make and forms the basis of most antenna types.

(ideally zero), and the maximum voltage also occurs at the outside ends. At the point in the center to which the transmitter is connected, we are a quarter wavelength back from the ends and thus the current is at a maximum, while the voltage is at a minimum. The magnetic field surrounding each increment of the current in an antenna will combine with the incremental electric field to form an electromagnetic field that leaves the dipole antenna primarily in a direction perpendicular to the axis of the wire, as shown in **Fig 16-2**.

The *electric field* (E) is parallel to the wire and defines the antenna's *polarization*. The *magnetic field* (H) is perpendicular to the wire axis. It's easy to tell the polarization of a dipole: If the wire is horizontal, the polarization is horizontal. Note, however, that horizontal and vertical have no particular meaning in space, except with reference to the Earth or some other planetary body!

The polarization is important since a horizontally polarized antenna will launch a horizontally polarized wave and vice versa. The reason it is important is that a horizontal antenna will also receive a horizontally polarized wave. Signals that travel via the ionosphere tend to have elements of both polarizations, due to random changes in the ionosphere. So having the same polarization is

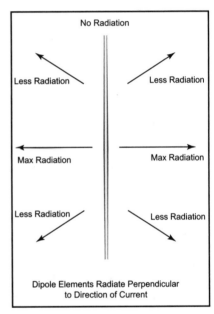

Fig 16-2—Direction of radiation from a dipole antenna.

usually only important for line-of-site or ground-wave signals, not signals that travel by means of the ionosphere.

In some point-to-point microwave systems scarce frequency resources are reused by transmitting two signals on the same frequency, one with horizontal polarization, and the other with vertical polarization. Very careful physical alignment of the antenna is required in order to keep the two signals separated.

More Complex Antenna Systems

It is possible to construct antenna systems using more than a simple dipole. Earlier we discussed how a reflector behind a light bulb makes the light appear stronger in some directions and weaker in others. Antennas work the same way and the energy leaving an antenna can be focused through an appropriate "lens" or reflector. A reflector, very much like that of a flashlight, can be effectively used at UHF and higher frequencies, where it can be of a manageable size. At lower frequencies it is more common to use less complex reflectors, as in **Fig 16-3**, or multiple-dipole elements[1] to focus the signals.

An important concept associated with radio waves is *phase*. This is simply a definition of the relative starting time of two waveforms of the same frequency. An electromagnetic wave will have a phase as well as an amplitude, and we can use both phase and amplitude to advantage. If two signals arrive at a receiving location in the same phase and at the same amplitude, they add together, resulting in twice as strong a signal. If the two equal-amplitude signals are 180° out-of-phase they will cancel. See **Fig 16-4**.

By careful placement of multiple antenna elements, along with the control of the phase of the signals in each, we can achieve many different forms of focused beams of electromagnetic energy. A simple example is shown in **Fig 16-5**, the *broadside array*. In this case two elements are used, although more can be combined to achieve a sharper focus. The array of Fig 16-5 can be placed in front of a reflector to result in a more sharply focused unidirectional signal.

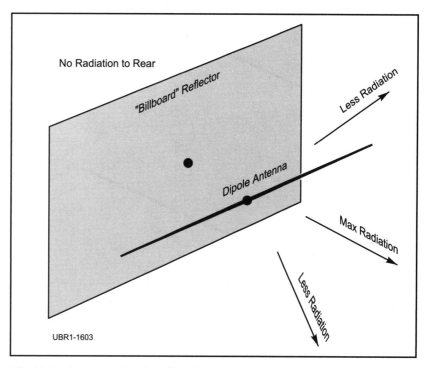

Fig 16-3—An example of a directional antenna—the *billboard* array.

It is also possible to form beams with dipoles not directly connected to a transmitter. These are called *parasitic arrays*. In a parasitic array, the unconnected but nearby, dipole picks up a portion of the signal from the directly connected antenna element and reradiates it to combine with the original signal in a way that meets the needs of the array.

It is important (and convenient) that all the properties of transmitting antennas extend to receiving antennas too. In other words, an antenna that transmits only in a particular direction will also receive solely from that direction. This makes unidirectional communications feasible with a single antenna.

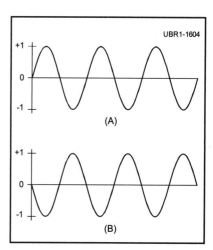

Fig 16-4—Two sine waves 180° out-of-phase.

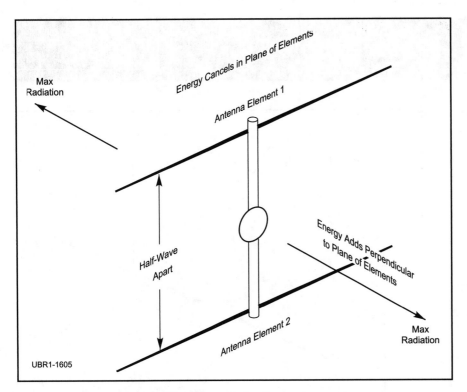

Fig 16-5—Energy focused through multiple dipole elements.

Effective Radiated Power

It is possible to consider (but difficult to construct) an antenna that radiates effectively in all directions. This is called an *isotropic* radiator. From the perspective of a receiving station, the signal from an isotropic radiator connected to a 1000 W transmitter appears the same strength as one from a 500 W transmitter connected to an antenna with 3 dB gain *in the direction of the* receiving station. A receiving station in another location, out of the focus of the transmitted beam may hear what sounds like a 10 W transmitter (20 dB below the peak gain), so whenever we discuss directive antennas, it is important to specify the direction toward which the gain applies.

In the direction of the beam, we can say that our 500 W transmitter results in 1000 W of *effective radiated power* or ERP. This is the called effective power because the receiving station receives a signal as strong as if it were a 1000 W transmitter radiating in all directions.

In some services, transmitter power is expressed in ERP and the operator can decide if it is better to have a larger transmitter and radiate in all directions, or have a lower-powered transmitter and focus the energy in particular directions. ERP can be expressed related to different reference power levels. For the case of comparison to an isotropic source, it is generally called EIRP with the *I* in this acronym indicating an isotropic reference.

The increase in effective radiated power from a transmitting antenna can be considered a gain, if you are in the right direction! I have placed gain in quotes in the heading because, unlike amplifier gain, you don't really have more power—you just have moved it around! A stronger signal in one direction occurs only because we have a weaker signal somewhere else. In many cases, this is exactly what you want, since you can avoid interfering with other users of the spectrum, and even may be able to avoid eaves-droppers.

The gain of an antenna in decibels (generally meaning the maximum gain in a particular direction) relative to an isotropic source is indicated as *dBi*, with the "i" referring to "isotropic." The dipole of Fig 16-1 in free space (with no ground reflections) can be shown to have a gain of about 2 dB relative to an isotropic source, or a gain of 2 dBi in the direction perpendicular to its axis. This is because there is no radiation in the direction of its ends and thus there is more radiation in the directions perpendicular to the dipole element.

Reference for a Directive Antenna

The dipole is also often used as a reference for comparison, since it is a well-defined configuration. It also can actually be built and is often used, although "free space" is a bit hard to find near the Earth's surface. Antenna gain referenced to a dipole in free space is indicated as *dBd*, and is *always* 2 dB less than dBi. This is important to keep in mind, since antenna literature quoting gain figures often doesn't make the distinction clear, and the result is a potential 2 dB confusion. The reflector antenna in Fig 16-3 might have a gain of 3 dB over a dipole in free space, but also a gain of 5 dB over an isotropic antenna in free space. Which number do you think

will be in the advertising brochure?

Simulation of a Directive Antenna

In Chapter 18 we will discuss antenna simulation in more detail, but as a preview, I will simulate two

antennas here using the antenna analysis tool *EZNEC*.[2] This will demonstrate some of the results we have been discussing. The first antenna is a dipole in free space as a reference. I picked a frequency of 30 MHz for no particular reason. The

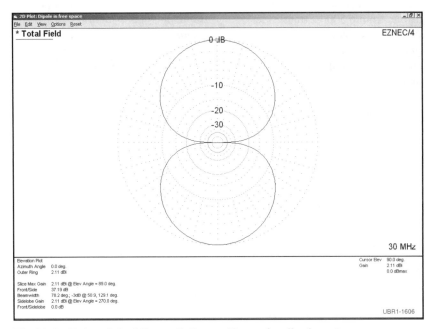

Fig 16-6—Polar plot of the radiation pattern of a dipole antenna.

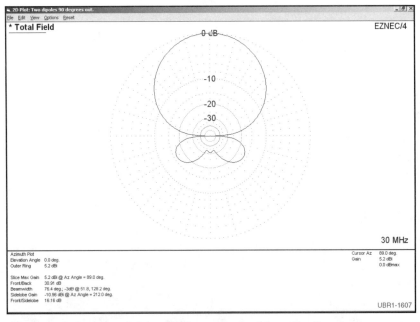

Fig 16-7—Polar plot of the radiation pattern of two dipole antennas spaced 1/4 λ and fed 90° out of phase to result in a unidirectional pattern.

Did you hear about the two antennas that got married —the ceremony wasn't great but the reception was wonderful!

radiation pattern in **Fig 16-6** would be the same for any frequency if I adjusted the antenna to be a half wavelength long. As shown in Fig 16-6 the maximum radiation is perpendicular to the dipole, with nulls at the ends. To make a simulation of a gain antenna, I used two identical dipoles spaced $1/4$ λ apart and applied half the power to each. In order to make a unidirectional antenna, I adjusted the phase of the rear dipole so it would be 90° behind the front one. When the signal travels the $1/4$ λ from the rear dipole towards the front one it is in-phase and adds to the energy going forward. Similarly when the signal from the front dipole reaches the rear dipole it will be 90° later and the signals will be 180° out of phase cancel in the rear direction. The resulting radiation pattern is shown in **Fig 16-7**.

Notice that the forward gain is indicated as 5.2 dBi, or almost exactly 3 dB greater than the indicated gain of the dipole in Fig 16-6. Note that the predicted signal to the rear is shown as being reduced by more than 30 dB compared to the forward going power. This is a reduction of a factor of 1000—very handy if you want to keep your signal away from some areas.

Table 16-1 provides a listing of some expected gains of typical antennas, assuming that they are properly designed and adjusted. Again, the gains apply to both transmitting and receiving antennas. The figures shown are representative of well-designed and constructed antennas. Don't assume that any given model that seems to fit will have exactly the figures shown, but if they don't you may want to check into what the designer might have done better.

Table 16-1

Typical gains of various antenna configurations. (Your mileage may vary!)

Antenna Type	Gain over dipole (dBd)	Gain over Isotropic (dBi)
Isotropic	−2	0
Dipole (free space)	0	2
Dipole over ground	6	8
Full-wave loop	2	4
Plane reflector	4	6
Corner reflector	8	10
Two-element dipole array	4	6
Three-element dipole array	6	8
Six-foot Parabolic dish antenna at 432 MHz	14	16
Six-foot Parabolic dish antenna at 1296 MHz	23	25

Notes

[1]Wolfgang, L, *Understanding Basic Electronics*, ARRL, Chapter 9. Available from the ARRL Bookstore. Order number 3983. Telephone toll-free in the US 888-277-5289, or 860-594-0355; **www.arrl.org/shop/ pubsales@arrl.org**.

[2]Basic and professional versions of *EZNEC* are available from Roy Lewallen, W7EL, at **www.eznec.com**. A scaled down version is supplied with *The ARRL Antenna Book*, 20th Edition, available from the ARRL Bookstore at **www.arrl.org/catalog/** order number 9043.

Review Questions

1. Compute the approximate length of a "half-wave dipole" antenna designed for the AM broadcast band, say 1.0 MHz; the 31-meter international broadcast band at 9.5 MHz; TV channel 2 at 58 MHz; and a satellite earth station downlink at 3000 MHz. What do you notice about the size of these antennas?

2. Under what conditions does a 500 W transmitter with an EIRP of 1000 W not appear the same as a 1000 W transmitter with an isotropic antenna?

Chapter 17

Receiving Signals, the Other Side of the Coin

Homemade 70-cm panel reflector.

Contents

In Chapter 16, I stated that the directional properties of a transmitting antenna also applied when it is used as a receiving antenna. Other parameters are applicable for both receiving and transmitting; however, there are some properties that are applied in a different way. You will need to understand the differences to be ready to deal with the system calculations later in this book.

Receive Anenna "Gain"

Chapter 16 discussed focusing the energy leaving an antenna by using a reflector, or by using combinations of elements driven with the appropriate phases. As noted then, a receiving station in the beam of such an antenna will receive a stronger signal than if the transmitting antenna worked equally well in all directions.

The difference is called the *gain* of the antenna, and works for reception as well as transmission. A stronger signal from one direction results because weaker signals are received from other directions. For a transmitting antenna, you can use the directional properties to foil eavesdroppers, or to avoid causing interference to users in other directions. With a receiving antenna you can eliminate some interference if it is coming from a different direction than your desired station.

Let's review some antenna parameters. **Fig 17-1** shows a uniformly radiating source, represented by a sphere. The radiation moves away from the source equally in all directions. This is known as an *isotropic source* and is often used as a reference. In practice, it is almost impossible to build one, so the isotropic source is a theoretical antenna.

The gain of an antenna in decibels (generally meaning the maximum gain in a particular direction) relative to an isotropic source is indicated in terms of *dBi*, with the "i" meaning

referenced to "isotropic." The dipole of **Fig 17-2** in free space (and thus with no ground reflections) has a gain of about 2 dB relative to an isotropic source, which is also a gain of 2 dBi. This gain is in the direction perpendicular to the dipole's axis. This gain occurs because there is no radiation in the directions from its ends and thus there is more signal in the directions perpendicular to the dipole.

Antenna Directivity

Antenna directivity is a term that is sometimes used in place of antenna gain, since any antenna with gain has directivity—the focusing of radiation in particular directions. The term antenna directivity should be used carefully, however. Some antennas

have directivity without gain, and some may even have less gain than an isotropic antenna. This can happen in two ways:

• An antenna can achieve directivity by having the energy that would be transmitted in undesired directions dissipated in a resistive load, rather than being redirected toward desired directions. Such antennas tend to be able to operate over a wide frequency range compared to gain antennas with frequency dependent elements.

• Some antennas, particularly those used only for receiving, can be very inefficient collectors of signal energy. This can be an acceptable compromise if they also only receive from a particular direction, and if the operating frequency is

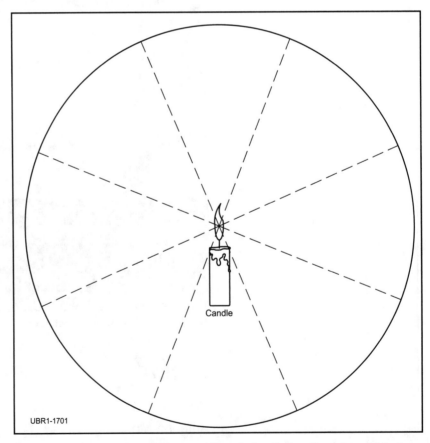

UBR1-1701

Fig 17-1—A source that radiates uniformly in all directions is said to be isotropic.

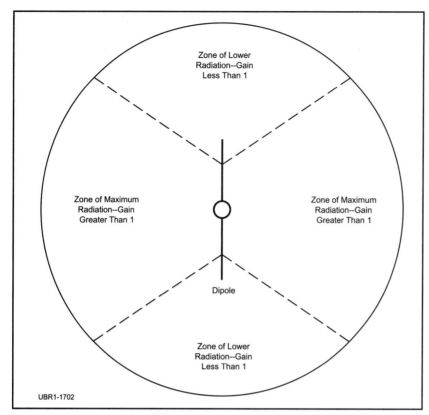

Fig 17-2—A dipole radiates primarily in the direction perpendicular to the wire making up the antenna. It has progressively less gain as the receiver moves in the direction of the ends. Note that the response in effect circles the antenna.

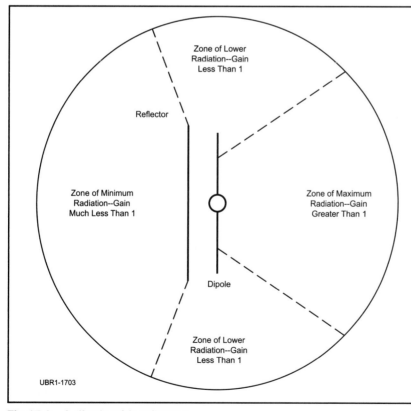

Fig 17-3—A dipole with reflector.

such that all noise is heavily dominated by external sources. This is often the case at lower HF and below. A good example is the internal loop antenna in an AM broadcast band radio. It picks up much less signal than a full-sized dipole, but also picks up atmospheric noise reduced by the same amount. The radio just has to have additional gain to make up for the lower signal levels.

Antenna Aperture

While antenna gain applies equally well to transmitting and receiving antennas, a related term, *antenna aperture,* shows up for receive only. In Chapter 15, I discussed a radiated field in terms of the power density, expressed in W/m^2. I then described the energy received in terms of the area of a receiving antenna. The area could be the size of an eye's pupil for light, or it could be the size of a receiving antenna for radio. If you know the field strength in W/m^2 and you know the effective area of the receiving structure in m^2, you can easily compute the received energy. The received power density times the receiving area equals the received power, as in $W/m^2 \times m^2 = W$.

So What's the Deal With Gain?

Well, this is where it may get confusing. You will notice that there were no gain factors in the above expression, although they're really buried in there. That is to say, the transmit power density will depend on distance, as well as transmitted power and antenna gain. The trick is to determine how to relate what we have been talking about to the *receiving area*.

Again, in radio terms, the receiving area is called the antenna aperture. This is the *effective size* of the receiving antenna. For some kinds of antennas, the effective size is related to the physical size in a clearly defined way. For example, in Table 17-1, a "six-foot diameter" parabolic reflector antenna is listed. If the six-foot diameter captures all the energy that comes its way, it

Table 17-1

Typical gains of various antenna configurations. Your mileage may vary.

Antenna Type	Gain over dipole (dBd)	Gain over Isotropic (dBi)
Isotropic	−1.8	0
Dipole (free space)	0	1.8
Dipole over ground	6	7.8
Full-wave loop	2	3.8
Plane reflector	4	5.8
Corner reflector	8.2	10
Two-element dipole array	4	5.8
Three-element dipole array	6	7.8
Six-foot Parabolic dish antenna at 432 MHz	14	15.8
Six-foot Parabolic dish antenna at 1296 MHz	23.2	25

would have an aperture of $\pi \times (6/2)^2$ or about 28 square feet, which is 2.6 m². The effective aperture will always be somewhat less due to feed design and losses, surface irregularities and other factors, but the usual antenna will gather up much of the energy that crosses its physical aperture. Now it's important to recognize that an antenna of constant physical size has increasing gain as the frequency increases.

Thus, the power density on the transmit side is a function only of antenna gain, while on the receive side the received energy is a function of antenna size. In terms of gain in a well-designed antenna system, a physically larger antenna will generally have more gain, and hence more aperture in receive. The concepts of gain and aperture are thus compatible, but these terms can cause some confusion in how we proceed to system analysis.

It is important to emphasize the relationship to frequency. If we have the same received power density (let's say, the same transmit power and antenna gain, at some fixed distance from the source), and we have the same effective receive antenna aperture—equally efficient—we will receive the same signal power no matter what the frequency. The difference is that as the frequency goes up, and a half-wave dipole gets physically smaller, for an antenna to have the same

aperture, it must have higher gain. It also follows that it will receive over a narrower angular range than at a lower frequency. Sometimes that's a benefit—sometimes not.

While it's easy to visualize the receive aperture of a parabolic dish-type microwave antenna, it is not so obvious that a very thin dipole antenna has any significant aperture at all. In fact, a dipole can be considered to respond to energy surrounding it for a particular distance around it. This can be determined in a number

of ways theoretically,[1] or perhaps most easily this can be shown by comparing the dipole's gain to that of an efficient dish antenna of a given size. By whatever method you use, the effective area of a half-wave dipole—to a radiated signal coming towards it perpendicular to the wire—can be shown to be an equivalent rectangle ½ wavelength long along the wire by ¼ wavelength perpendicular to the approaching wave front. Note that while the rectangular-shaped figure is easy to calculate,

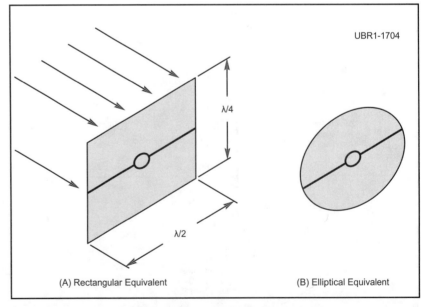

(A) Rectangular Equivalent   (B) Elliptical Equivalent

Fig 17-4—The effective aperture of a dipole antenna. At A, the rectangular equivalent. At B, another area, perhaps more accurately showing the importance of different parts of the antenna.

other shapes with an area of $\lambda^2/8$ may really be more accurate representations of the importance of different parts of the antenna.

Two possibilities are illustrated in **Fig 17-4**.[2]

A way to experimentally verify that this is a real effect is to have two dipoles in the field of, and broadside to, an incoming wave. If they are a few wavelengths apart, they should each receive the same signal strength, as if the other dipole didn't exist. As they get within their effective aperture of each other, the signal energy starts to become shared between them, and the signal received by each is reduced. At some point each will receive half the power that they received when separated from each other.

The standard reference isotropic antenna can be shown to have an effective aperture of $\lambda^2/4\,\pi$. Note that if you take the aperture ratio of the isotropic to the dipole, you get $(\lambda^2/8)/(\lambda^2/4\,\pi)$ or $4\,\pi/8$. This equals a power ratio of 1.57, which is 1.96 dB. This is what you would expect for the gain of a dipole over an isotropic source. You can generalize by saying that if you know the gain of an antenna, you can determine the effective aperture as follows:

$$A_{eff} = G_i \lambda^2/4\,\pi \qquad \text{(Eq 17-1)}$$

where G_i is the power gain compared to an isotropic antenna. This is also:

$$A_{eff} = G_d\,\lambda^2/8 \qquad \text{(Eq 17-2)}$$

where G_d is the power gain compared to a dipole antenna.

What this all means is that to get the same received signal level at two frequencies, the receive antenna needs to be about the same "size," all other things being equal. If you calculate received power based on antenna gain, it will appear that lower frequencies have an advantage and that there is a kind of loss associated with higher frequencies. While this representation is internally consistent and can be used, it is important to remember that higher frequencies don't really have more loss, just smaller antennas for the same gain. I hope that makes sense. I'll take another stab at this when I discuss path loss in the next chapter!

Antenna Modeling

You may well wonder how it is that you can determine just how well and in which directions an antenna will radiate. This has been an age-old problem and traditionally was best settled on an *antenna range*. This is a large open space with appropriate towers and measuring equipment so that the actual directivity and gain of an antenna can be measured. Of course, there have always been theoretical methods to determine antenna performance, but they have often suffered from a lack of inclusion of all of the important factors. Key elements excluded have often been the effects of ground reflections and antenna interaction with other nearby objects, such as support structures, or vehicle structure for mobile, airborne or shipboard antennas. Fortunately, in our twenty-first century computer-oriented society, we have software tools that help avoid the need to climb towers!

The performance of an antenna can be computed by dividing up a model of an antenna into a number of wire *segments*, and adding up the radiation coming from each of the segments, considering both magnitude and phase and the mutual interaction with other segments. The idea behind segmentation into small increments is that each is small enough so that the current leaving that segment is the same as the current entering it. The program thus determines the effects of each "mini antenna" at a distant location. Other structures can be modeled also, such as metal towers, guy wires and other antennas not actually connected to

the antenna being "tested." The signals induced in these other nearby conductors will be part of the final calculation.

The result can be an effective tool to predict how a particular antenna will function, without having to actually construct and measure the way it really works. As with any simulation, your mileage may vary, but under most circumstances modeling provides at least a good starting point.

EZNEC

One popular antenna-modeling program is named *EZNEC* (pronounced *easy-neck)*. The "NEC" part of the name comes from the core calculation engine, *Numerical Electromagnetics Code*, a powerful antenna analysis tool that forms the basis of a number of antenna analysis programs. The "EZ" part of the name comes from the fact that this *Windows* implementation is easy to use.

To use *EZNEC*, you start with an antenna definition file (some sample antenna models are supplied with the program). You enter in the physical dimensions of your antenna in X, Y and Z coordinates, specifying wire size or diameter and picking a segment quantity (the basic version has a limit of 500 segments) on the WIRES tab. On the SOURCES tab you indicate which wires will be connected to sources, and the location where the source connection will go in terms of percentage distance from one end of the wire. Pick the type of ground you want from the GROUND tab choices.

That's all it takes to model an antenna. To determine the results, you may select the SWR tab and give the program a range of frequencies over which to determine the antenna's impedance and SWR (default is 50 Ω, but you can select other impedances). If you want a plot of the antenna pattern, you can select azimuth, elevation or even 3-D plots on the PLOT TYPE tab and then hit the FF PLOT (far-field plot) to see the results. It takes less time to do it than talk about it!

If you want to find out the effect of changing the length, you can adjust the dimensions in the WIRES tab and run it again. Interested in finding out how it works at another frequency? Just enter another frequency in the FREQUENCY tab and hit FF PLOT again.

Of course, there are some refinements and details that make *EZNEC* even more useful, but I expect you will find out about them as you need them. We will go through some examples in the next chapter.

Notes

[1]J. Kraus, W8JK (silent key), *Antennas*, First Edition, 1950, McGraw-Hill Book Company, NY, 1950, pp 50-54.

[2]Ibid, Figure 3-7, p 52.

[3]Basic and professional versions of *EZNEC* are available from Roy Lewallen, W7EL, at **www.eznec.com**. A scaled down version is supplied with *The ARRL Antenna Book*, 20th Edition, available from the ARRL Bookstore at **www.arrl.org/catalog/** order number 9043.

Review Questions

1. Discuss an application in which a transmitting antenna with isotropic directivity might be beneficial. Repeat for a highly directive antenna.

2. What is the maximum expected aperture of a dish antenna with a 10-foot diameter? Repeat for a half-wave resonant dipole antenna at 5.34 MHz. Discuss the results.

3. Discuss precautions needed to make sure a modeled antenna will have results that will predict real-world antenna performance.

Chapter 18

Using EZNEC to Model Real Antennas

It's Field Day 2004 and some hams are testing their satellite antennas!

Contents

I have put together some *EZNEC* models of some simple antennas to serve two purposes. First, to show how easy it is to use, and second to illustrate some antenna principles that I discussed in an abstract way earlier.

Dipole In Free Space

The first antenna I will model is a horizontal half-wave dipole in free space. I picked a frequency of 10.0 MHz (30 meters) for no particular reason. **Fig 18-1** is the **Wires** entry table that I made. Note that I have picked this length so that the antenna is resonant at 10.0 MHz. While resonance isn't mandatory in an antenna, this truly represents a "half-wave resonant dipole."

I selected a height (the Z coordinate) of half a wavelength above ground, which is 49.2 feet at 10.0 MHz. Note that in "free space" a ground height doesn't matter. But you must tell *EZNEC* something—I could just as well have chosen zero feet for free space, but I knew that later in the modeling session I would change the ground type to "real ground."

Next, I specified AWG #14 wire for the diameter. I could have specified this as a wire size (using the # symbol) or entered the diameter in inches, or mm, since you can set the table up in metric or English units.

I selected a number of segments (51) rather arbitrarily. The program

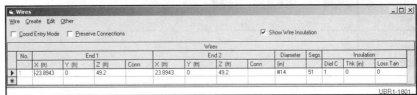

Fig 18-1—*EZNEC* Wires entry for a 30-meter dipole model.

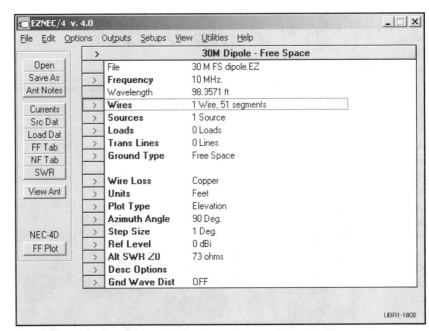

Fig 18-2—Main *EZNEC* screen for a 30-meter dipole model.

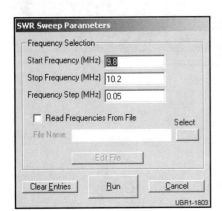

Fig 18-3—Entry window for SWR measurements.

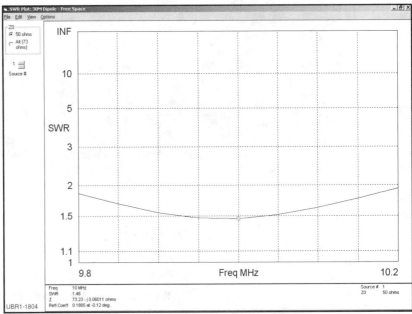

Fig 18-4—SWR measurement results for dipole, with cursor placed at 10.0 MHz.

will give a warning message if you select a number of segments that result in too small or too large a segment length. Keeping in mind the 500 segment limit in *EZNEC*, one way to gauge whether the segment size is appropriate is to temporarily double the number and see if the result changes very much. If it doesn't, then your original segment number is likely to provide valid results.

Now we've got a model, and the *EZNEC* main screen should look like **Fig 18-2**. By clicking the **SWR** button, you bring up the screen in **Fig 18-3** on which you can specify the range and resolution you would like for an SWR plot. **Fig 18-4** shows the plot examining the SWR around the resonant frequency. Note that by clicking the cursor on the frequency axis it shows the numerical data, here, for 10 MHz. You could also select any other frequency in the range. Note also that I used this screen to "trim" the antenna to be resonant (reactance close to zero ohms). I made the dimensions longer if the reactance at 10.0 MHz were negative, and shorter for a positive reactance, until I got quite close to zero. Note also that my final length, 29.8929 feet on each side, is a level of precision that works in the model, but is not realistic for a ruler and wire cutters. A tenth of an inch is asking a lot in the real world! The message is, the model is fine as far as it goes, but don't let it result in unrealizable situations.

If I were actually making an antenna like this (and I have many times!), I would use the model as a guide and start with perhaps an extra six inches of antenna length on each side, raise the antenna to its operating height temporarily, measure the SWR over the desired frequency range, and then lower the antenna and shorten the ends a little at a time until I had the results I wanted.

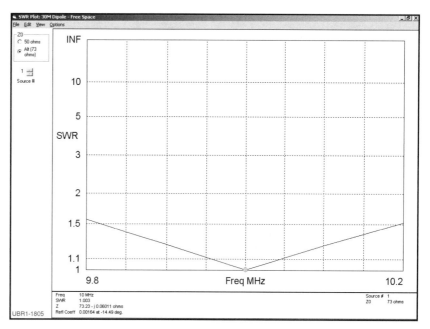

Fig 18-5—SWR measurement results using alternate Z_0 of 73 Ω.

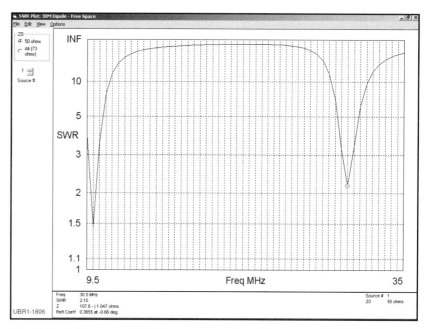

Fig 18-6—Wide range SWR measurements, with cursor at 30.5 MHz.

If you wanted to adjust the model to indicate the SWR at a different Z_0, you merely change the **Alt SWR Z_0** on the main *EZNEC* screen, for example to the nominal 73 Ω that the antenna shows at resonance. See **Fig 18-5**. You could also use the model to calculate the SWR over a wider frequency range. Just make another pass and change the SWR entry frequency limits. The results are shown in **Fig 18-6**. This highlights the fact that most antenna types have multiple resonances, as seen at slightly above the third harmonic of the 10 MHz 3/2-wave, "third harmonic" resonance at 30.5 MHz. Some antennas take advantage of this effect. For example, hams often use a 40-meter dipole on its third harmonic resonance at 15 meters.

The antenna patterns of the free-space dipole are shown in **Fig 18-7** (azimuth) and **Fig 18-8** (elevation). The length of a radial line drawn from the center of the plot to the pattern at a particular angle determines the signal strength at that angle, relative to the peak level of the pattern. In the case of a dipole in free space, the elevation pattern shown in Fig 18-8 is constant all the way around the antenna.

The elevation pattern shown for the free-space dipole in Fig 18-7 is typical of what we would expect of a horizontal dipole in free space—the typical "figure-8" pattern with maximum radiation broadside to the antenna and no radiation from the ends. As expected, the maximum broadside radiation is about 2 dBi.

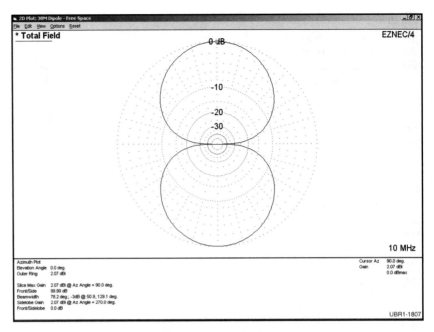

Fig 18-7—Azimuth pattern of dipole model in free space.

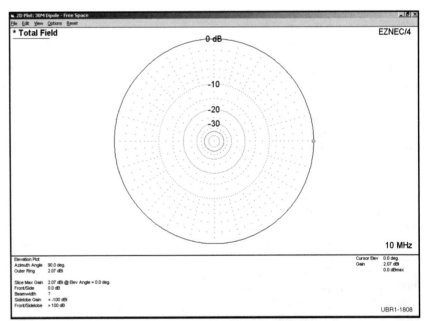

Fig 18-8—Elevation pattern of dipole model in free space.

The previous section dealt with an ideal dipole in free space. While space is still pretty cheap, free space is still really hard to find. Most dipoles are fabricated closer to the earth! A feature of *EZNEC* is the ability to easily move between theoretical free space, theoretically perfect ground (maybe an infinite sheet of perfectly conducting gold) and real ground. By just clicking on the **Ground Type** button on the main *EZNEC* screen, you are offered a choice. I have chosen "Real/High Accuracy" in **Fig 18-9**. You can accept the EZNEC default ground parameters of "Medium (0.005, 13)" or insert actual parameters depending on your soil conditions.

In **Fig 18-10**, you see the resulting SWR plot for this dipole over real ground. While it has the same shape as the free space plot, notice that the resonant frequency has shifted higher. This is a typical result following the change of most anything about an antenna and it's one reason why many antennas have adjustable element lengths—to compensate for real-world conditions.

The elevation plot in **Fig 18-11** is very different from the same plot for the free-space antenna. While Fig 18-8 was uniform all around the antenna, Fig 18-11 dramatically shows the result of a reflected signal from the earth below. For a horizontal antenna, the reflection is out-of-phase with the incident wave. At the half wave height specified, the radiation reflected from the ground reinforces the direct wave going upward from the antenna at an elevation angle of 28° with respect to the flat ground. Similarly the direct wave and the out-of-phase reflected wave cancel at the horizon (0° elevation). In general, as the antenna is raised higher above the earth, the main radiation lobe is lowered, but it never quite reaches the horizon.

Fig 18-12 provides a plot of the azimuth pattern at a 28° elevation angle. Note that *EZNEC* allows you to specify any elevation angle for the azimuth plot. I selected the elevation with the maximum signal, but you may have interest in the signal at other angles. In this plot, *EZNEC* has selected the azimuth with the strongest signal to show the detailed data at the bottom of the plot. As

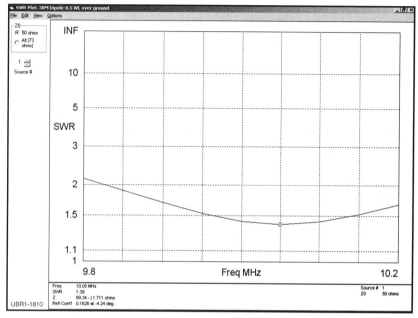

Fig 18-9—SWR of antenna over real ground. Note shift in resonant frequency compared to Fig 18-5.

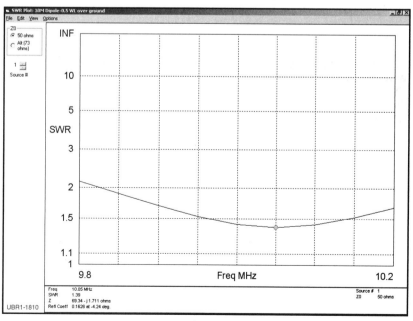

Fig 18-10—Free-space model changed to be over "real ground."

Fig 18-11—Elevation plot of antenna over real ground. Compare to free-space elevation plot in Fig 18-8. At 28° elevation, the gain is 5.4 dB higher than the free-space antenna, due to the ground-reflection effect.

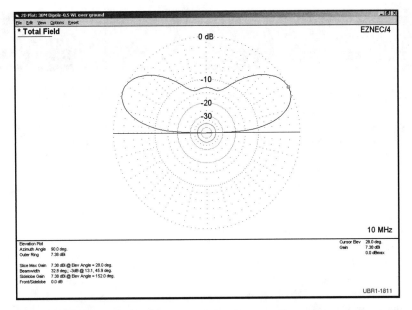

Fig 18-12—Azimuth plot of dipole antenna ½ λ over real ground.

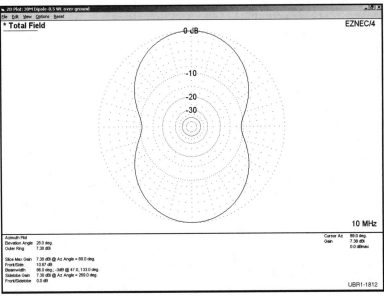

Fig 18-13—Azimuth plot of antenna ½ λ over real ground. In this view the cursor was moved to 45° off the main beam, to show the gain at that point.

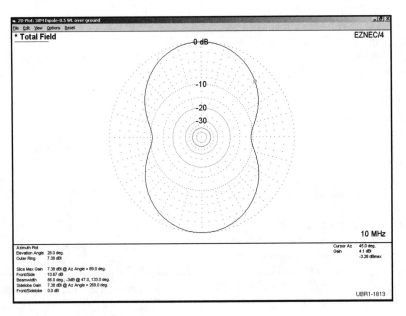

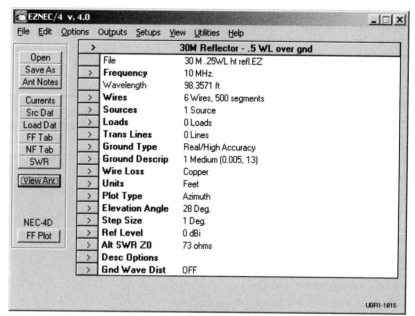

Fig 18-14—Wires table for simulated screen-reflector antenna.

Wires (menu: Wire, Create, Edit, Other)

☐ Coord Entry Mode ☐ Preserve Connections ☑ Show Wire Insulation

No.	End 1				End 2				Diameter	Segs	Insulation		
	X (ft)	Y (ft)	Z (ft)	Conn	X (ft)	Y (ft)	Z (ft)	Conn	(in)		Diel C	Thk (in)	Loss Tan
1	-23.8929	0	49.2		23.8929	0	49.2		#14	101	1	0	0
2	-36	-24.6	49.2		36	-24.6	49.2		#14	99	1	0	0
3	-36	-24.6	46.2		36	-24.6	46.2		#14	99	1	0	0
4	-36	-24.6	52.2		36	-24.6	52.2		#14	99	1	0	0
5	-36	-24.6	42.2		36	-24.6	42.2		#14	51	1	0	0
6	-36	-24.6	56.2		36	-24.6	56.2		#14	51	1	0	0

with all *EZNEC* plots, merely clicking on another point in the plot shows detailed numeric data for that point. In **Fig 18-13**, I have selected the data for 45° off axis, just by clicking on that point of the curve. This is a very powerful tool!

More Complicated Models

It is easy to add in other conducting objects, such as electrical wiring, pipes or other antennas. Usually the segment limit of 500 in the basic version determines how much you can actually insert into your model. The *EZNEC pro* version can handle up to 20,000 segments, but it costs significantly more than basic version.

As an example, I have attempted to model the "billboard antenna" shown in Fig 11-13 in Chapter 11 by using a number of rods in place of the solid screen shown. This is a fairly common implementation at lower frequencies, where a solid reflecting plate would be impractical. **Fig 18-14** is the **Wires** table of a five-wire implementation of such a screen reflector array. I have made each wire 50% longer than the antenna element and spaced them vertically above and below the antenna center at a spacing of ¹/₄ λ from the antenna.

Fig 18-15 shows the new *EZNEC* main screen. In case you didn't count them up, it shows that I have just hit the 500 segment limit, and I used fewer segments on the most distant rods. As you might expect by now, the SWR plot in **Fig 18-16** for this dipole plus reflector array is different than it is for the dipole without a reflector. The resonant frequency has

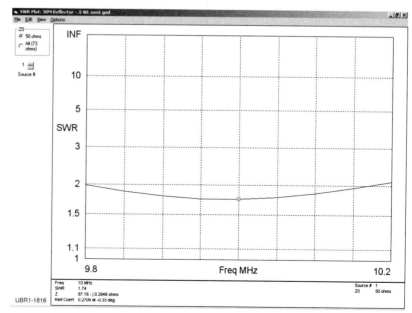

Fig 18-15—Main *EZNEC* screen for screen-reflector antenna. Note that the maximum of 500 segments is now in use.

Fig 18-16—SWR plot of screen-reflector antenna. The impedance is different, typical of most antenna changes.

shifted back down and the impedance at resonance is higher.

Fig 18-17 shows the elevation pattern. Note that unlike the dipole, which had a symmetrical pattern between front to back, this antenna has a definite front that's different from the back. In fact there is about 2 dB more signal going towards the front than from the dipole by itself, and a lot less going to the back.
Fig 18-18 shows the azimuth pattern and indicates on the left side of the data that we do indeed have a "front-to-back ratio" of 7.44 dB. This is just what we would like from such an antenna—with more of the signal going where we want it, and less going where we don't. I didn't spend any time trying to optimize this antenna; however, I would expect a real billboard antenna to have a front to back ratio closer to 30 dB and a gain of at least 3 dB, so that if I could have modeled more wires it would be closer to a real solid reflector.

It's also important to note that a single wire about 5% longer than the antenna element and spaced about 0.15 λ behind it would result in far better forward gain and front-to-back ratio than our simulated billboard at the design frequency. I'll leave that as an exercise for the student. The billboard, on the other hand, will work well over a wide range of frequencies, through and including the second resonance, while the resonant parasitic reflector formed by the single wire will provide directivity only over a relatively narrow frequency range. Each type of antenna has its own most appropriate applications.

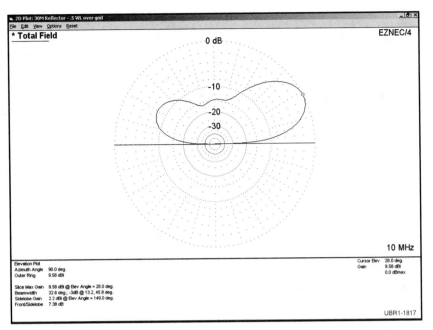

Fig 18-17—Elevation plot of screen-reflector antenna. Note that the front lobe is more than 2 dB stronger than that from the dipole in Fig 18-11.

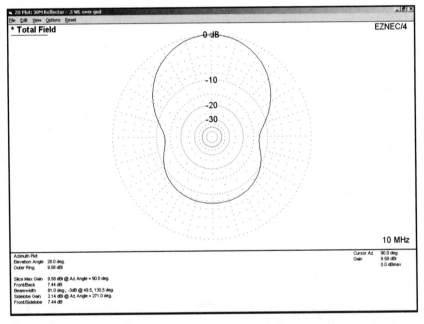

Fig 18-18—Azimuth plot of reflector antenna. The front-to-back ratio is 7.32 dB.

As I mentioned previously, usually the antenna and radio are not in exactly the same place. There are some notable exceptions, particularly in portable hand-held systems and various microwave communications and radar systems. But in most other cases, optimum performance requires the transmitter and receiver to be located some distance from the antenna.

The component that makes the connection between the transmitter or receiver and its antenna is called a *transmission line*. Transmission lines are used in places besides radio systems—for example, power distribution lines are a kind of transmission line, as are telephone wires and cable TV connections.

In addition to just transporting signals, transmission lines have some important properties that we will need to understand to allow us to make proper use of them. This section will briefly discuss the key parameters.

Characteristic Impedance

A transmission line generally is composed of two conductors, either parallel wires such as we see on power transmission poles, or one wire surrounding the other, as in coaxial cable TV wire. The two configurations are shown in **Fig 18-19**. Either type has a certain inductance and capacitance per unit length and can be modeled as shown in **Fig 18-20**, with the values determined by the physical dimensions and properties of the insulating material between the conductors.

If a voltage or signal is applied to such a network, there will be an initial current flow independent of what's on the far end of the line, but based only on the L and C values. The initial current will be the result of the source charging the shunt capacitors through the series inductors and will be the same as if the source were connected to a resistor

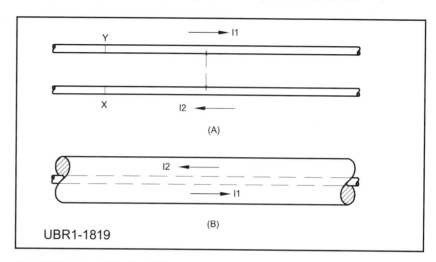

Fig 18-19—Parallel wire (A) and coaxial transmission lines (B).

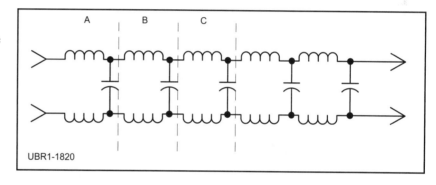

Fig 18-20—Lumped constant equivalent of an ideal transmission line.

whose value is equal to the square root of L/C.

If the far end of the line is terminated in a resistive load of the same value as the resistor above, all the power sent down the line will be delivered to the load. This is called a *matched* condition. The value of the resistor in this scenario is called the *characteristic impedance* of the transmission line, and is perhaps the most important parameter associated a transmission line. Common coaxial transmission lines have characteristic impedances (referred to as Z_0) between 50 and 100 Ω, while balanced, two-wire transmission lines are found with characteristic impedances in the range of 70 to

600 Ω. What this means to us as radio people, is that if we have an antenna that has an impedance of 50 Ω and a radio transmitter designed to drive a 50 Ω load, we can connect the two with any length of 50 Ω coaxial cable, and the transmitter will think it is right next to the antenna. The antenna will receive most (see next section) of the transmitted power and all is well with the world!

Attenuation

The ideal transmission line model shown in Fig 18-20 passes all input power to a matched load at the output. A real transmission line also has loss resistance associated with the wire conductors and some loss of signal due to the nature of the insulating material. As transmission lines are made larger, the resistance is reduced, and as the dielectric material gets closer to low-loss air, the dielectric losses are reduced. The skin effect causes currents to travel nearer to the surface of the conductors at higher frequencies, and the effective loss thus increases as the frequency is increased.

Fig 18-21 provides some real world examples of the losses as a function of frequency for the most common types of transmission line. Note that the loss increases linearly with length and the values are for a length of 100 feet. The losses shown are for transmission lines feeding loads matched to their Z_0. As will be discussed shortly, losses can increase significantly if the line is not matched.

The "open-wire" line shown in Fig 18-19A consists of two parallel wires with air dielectric and infrequent spacers. Such a line typically has a characteristic impedance of 600 Ω. While the losses of such a line are low, they only work well if spaced away from metal objects and if they are not coiled up. While coaxial cables have higher inherent losses, all the signal is kept within the outer shield conductor. Coax cables can be run in conduit, coiled up, placed next to other wires and so are much more convenient to work with. Sometimes a long straight run of low-loss open-wire line will be transformed to 50 Ω at the each end, with coaxial cable connected to the antenna end and to the radio end to take advantage of the benefits of both types of transmission lines.

Velocity of Propagation

Signals in air-dielectric transmission lines propagate at almost the speed of light. Other dielectric materials cause the signals traveling in transmission lines to slow down, just as we observe with light rays traveling through water. In many cases, this is not a matter of concern, since we often only care that the signals get out the other end. However, there are some exceptions.

The velocity can be shown to be reduced by a factor of one over the square root of the dielectric constant relative to air. Some cable specifications provide the relative velocity as a fraction of the speed of light. Most engineering handbooks include tables of properties of materials. For example, polyethylene is a common cable insulating material and has a relative dielectric constant of 2.26.

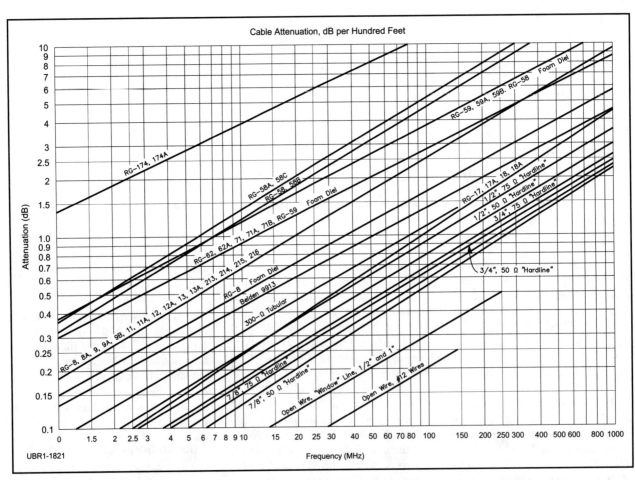

Fig 18-21—Loss of some typical transmission lines, in dB per 100 feet, as a function of frequency. The RG-58 transmission lines are 50 Ω polyethylene insulated coaxial cable slightly less than ¼ inch in diameter. The RG-8 through RG-216 lines are 50 and 70 Ω polyethylene insulated transmission lines with a diameter somewhat less than ½ inch. The "hardline" types have a foam dielectric that has properties close to that of air.

The square root of 2.26 is 1.5, so the propagation velocity in polyethylene insulated coaxial cable is $3/1.5 \times 10^8 = 2 \times 10^8$ m/sec.

Some applications actually use coaxial cables to provide delayed signals in pulse applications, and having a way to accurately predict the delay just by knowing the cable characteristics and measuring the length of the cable can save a lot of lab time. In the radio world, I have discussed previously about driving antenna elements in a particular phase relationship to obtain a desired antenna pattern. If a transmission line is used to provide the two signals of different phase, we need to know how fast the signal propagates in order to determine the effective line length.

As we will discuss in the next section, transmission lines of particular electrical lengths can be used as impedance transformers. However, unless we know the propagation velocity we can't determine the proper length.

Lines With Unmatched Terminations

So far I have discussed transmission lines feeding terminations matched to their characteristic impedance. If that is *not* the case— that is, there is a *mismatch* between the characteristic impedance of the transmission line and the load at its end— the voltage-current relationship at the load will reflect the impedance of the load—not the characteristic impedance of the line. Further, along the line the voltage and current will vary with distance from the load. Thus the impedance presented to the transmitter end will be neither that of the far end Z_L, nor the Z_0 of the transmission line itself. The impedance at the transmitter end of the line can be calculated, knowing the Z_L, the Z_0 and the electrical characteristics of the line, including its length.

The ratio of maximum voltage on the line to minimum voltage on the line is called the *standing wave ratio* or SWR. A matched line has an SWR of 1:1. A 50-Ω line terminated with a 25 or 100-Ω load will have an SWR of 2:1. There is a whole family

of complex impedances that will also have a 2:1 SWR, by the way, but the computation is easier with resistive loads.

There are some interesting special cases with a mismatched line. For example, the load impedance, resistive or complex, repeats every $1/2\ \lambda$ along the line. The impedance goes to the opposite extreme at odd multiples of a $1/4\ \lambda$. For example, a 25-Ω load would get transformed to 100 Ω in a $1/4$ or a $3/4\ \lambda$ length of 50-Ω transmission line. This effect can be used to our advantage if we wish to transform impedances at a specific frequency.

A generally less desirable effect of mismatched lines, however, is that the losses increase. This is easy to understand. If the voltages and currents along the line are higher because of a mismatch between the load and the line's characteristic impedance, you can expect losses to increase as well. **Fig 18-22** shows the

additional loss for a mismatched line that needs to be added to the matched loss in Fig 18-21. As is evident, the combination of matched loss and high SWR can result in dramatic increases in overall loss. This is why antenna designs that don't use matched transmission lines often use air-dielectric lines that have inherently low matched-line losses.

Notes

[1] J. Kraus, W8JK (silent key), *Antennas*, First Edition, 1950, McGraw-Hill Book Company, NY, 1950, pp 50-54.
[2] Ibid, Figure 3-7, p 52.
[3] Basic and professional versions of *EZNEC* are available from Roy Lewallen, W7EL, at **www.eznec.com**. A scaled down version is supplied with *The ARRL Antenna Book*, 20th Edition, available from the ARRL Bookstore at **www.arrl.org/catalog/** order number 9043.

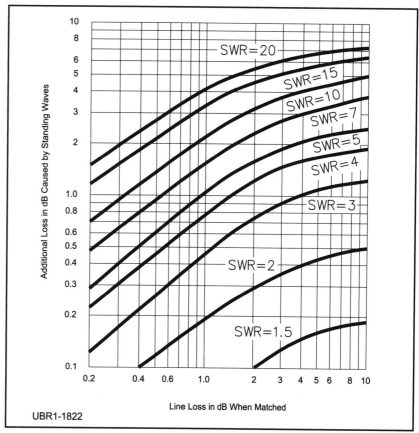

UBR1-1822

Fig 18-22—Additional loss of a transmission line when mismatched. This loss needs to be added to the matched-line loss in Fig 18-21 for mismatched lines.

Review Questions

1. Discuss the difference in the behavior of antennas near the ground and those in free space. Why must we understand about the ground conditions to make a usable model?

2. If you want to make a 1/4 λ section of RG-213 at 10 MHz how long would you make it, assuming that it has a velocity of propagation of 66% of that in air?

3. A 1000 W transmitter at 15 MHz is feeding a matched load through 200 feet of 50-Ω RG-8 transmission line. How much power reaches the antenna? Repeat if the frequency is 150 MHz. Repeat both cases if the antenna has an SWR of 3:1. [Hint: Look at Figs 18-21 and 18-22.]

Chapter 19

Line-of Sight Communications Systems

An elaborate VHF/UHF/microwave "rover" station put together by NØDQS on his 1983 Chevy Blazer for contesting via line-of-sight. Notice the PVC pipe-frame mounting system.

Contents

In the previous chapters, we have been talking about the various pieces and processes of radio systems. A good understanding of each element is important to make proper use of them. The next step is to put them together into a *system* that will meet a set of user requirements. As an individual who has spent decades as a telecommunications and radar systems engineer, I have to believe that this is "where the rubber meets the road," to quote an old advertising line.

The good news is that we don't have to be systems engineers to understand the issues, but with the knowledge we've gathered in the preceding chapters, we will be able to understand how the pieces fit together in order to accomplish the job we need. In this chapter, we will cover line-of-sight communications systems. In the next, we will extend the discussion to longer distance communication systems.

A Place to Start—Line-of-Sight Radio Links

A very frequently encountered and important radio system configuration is that making use of a *line-of-sight* (LOS) radio link. What we generally mean by a LOS link is a straight-line optical path (you could hit the destination with a searchlight, for example). This kind of path is not influenced by ionospheric propagation effects, except when operating earth-to-space communications. And even then, this must be at frequencies above the critical frequency, discussed in Chapter 15, to get through the ionosphere. Once we complete our discussion of LOS links, we will move to longer distance systems. All the elements of LOS systems apply to the other types; they just add in additional complications.

The analysis of an LOS system involves adding up all the gains and losses, determining the resultant

power delivered to the receiver and then comparing it to the combination of the received noise power and the internal receiver noise power. The result is a *signal-to-noise ratio*, usually abbreviated SNR. Each type of service has an SNR requirement for satisfactory operation, although it is often possible to use degraded operational capability at a lower SNR, especially during emergency situations. Typical examples of SNR requirements are shown in **Table 19-1**.

So How Can we Find Out if We're Line-of-Sight?

The analysis of LOS systems implies that there is a clear path between the two end points. Sometimes, it can be obvious—you go to one end-point and look towards the other end-point. If you can see it, it's a LOS path! In other cases, there may be a need for a way to predict the distance without being there, or because of visual obstacles. The most straightforward case is the one restricted by the curvature of the earth. This would apply to an over-water path or a path over flat land without obstructions.

The radio horizon and even the visual horizon are actually farther than we would think. Wave propagation through the atmosphere results

in refraction that is more significant the closer we are to the earth. This results in a bending of the path to extend it beyond the geometric horizon, one of those times that nature helps us! The additional distance in most parts of the world is about the same as if the earth were $4/3$ the size of the actual earth. **Fig 19-1** shows the geometry and defines the relationships.

If we don't have a smooth surface to deal with, either due to natural surface irregularities or other objects, we must clear those obstacles as well as the natural horizon. In most areas of the world, there are topological maps indicating ground contours. They may need to be appended with building construction details. It also should be kept in mind that if you don't own the land, or at least the air rights, you may want to keep your eye on construction activity after the link goes in!

It's not enough that the signal just clears the surface, or the obstruction. If it just cleared, the signal would refract or reflect in strange ways resulting in potentially destructive interference and not as much signal would reach the receiver, as we would expect. To have enough space so that the signal is not impacted, we should design to leave a vertical distance above the obstruction at

Table 19-1

Typical system minimal signal-to-noise ratio (SNR) requirements[1].

Signal type	Typical Minimum SNR
Relatively snow-free analog television signal	50 dB
Toll quality voice telephone service	30 dB
Tactical voice communication	12 dB
Non-error corrected data system	30 dB
Data system including strong error correction	10 dB
Morse telegraphy with highly trained operator	0 dB

[1]These are not hard and fast rules, but are based on the author's experience. Each individual system should document its specific operational requirements.

least equal to $\sqrt{\lambda d_1 d_2 / d}$, where d_1 is the distance from end one to the obstruction, d_2 is the distance from end two and d is the total path length, all in the same units as the wavelength. **Fig 19-2** shows the geometry involved when there is an obstruction to the line of sight path between two stations.

Adding up the Pieces

Fig 19-3 shows a block diagram of the elements in a radio communications system, while **Fig 19-4** shows key parameters in the relationships between the elements in a communication system. The key parameters are described below:

- Transmitter power output—this is the average or peak power (depending on system type) leaving the transmitter.
- Feed line loss—In general the transmitter is located some distance from the antenna. The transmission system has finite loss and this will reduce the power delivered to or from the antenna.
- Antenna gain—The antenna design may focus the energy in the direction of the receiver. If so, the effect is to increase the signal launched towards the receiver. In addition, ground reflection may add or subtract from the signal leaving at the desired take-off angle.
- Effective radiated power (ERP)—This is a parameter that combines the previous three. As noted in the previous chapter, a receiving system can't tell the difference between a 100 W signal from an isotropic source and a 50 W signal transmitted from a system with a 3 dB gain antenna pointed at the receiver. Thus from the point of view of the system design, it is all the same to the far end, whichever is deployed. In other words, the signal that enters the path towards the receiver can be generated by high power and a low gain antenna or low power and a high gain antenna and have the same effect *if the antenna is accurately pointed.* Thus the system design can specify a required ERP and let the transmitter site designer decide how to

generate the needed ERP.
- Path loss—The signal loss between transmitter and receiver can include a number of factors. First, for a LOS system, is the reduction in signal per unit area based on path distance, as will be described later. In addition, there may be other factors that tend to reduce the signal, some constant and some variable. Constant attenuation factors include obstacles in the path such as buildings, other structures and vegetation as well as the effect of reflections that can add or subtract from the direct signal. Variable

attenuation factors include rainfall, birds and aircraft flying through the path. Sometimes wind deflection of antennas and other mechanisms result in signal fluctuation or *fade.* Note that some of the factors come into play at different frequencies. For example, rainwater absorbs a significant amount of radiation in the 12 GHz band, and almost none at HF.
- Antenna aperture—The signal that is intercepted by the receive antenna aperture is converted from a wave front to an electrical signal on the feed line. As noted previously, the effective aperture can be

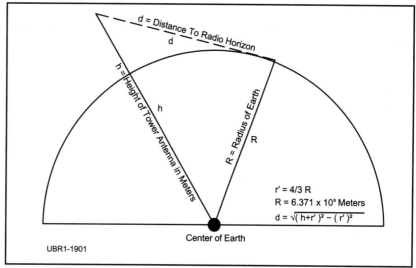

Fig 19-1—Distance to the horizon for a ⁴/₃-Earth radius line-of-sight path.

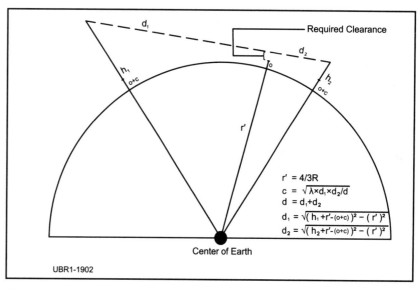

Fig 19-2—Distance to the horizon for a ⁴/₃-Earth radius line-of-sight path, but this time with an obstruction in the way.

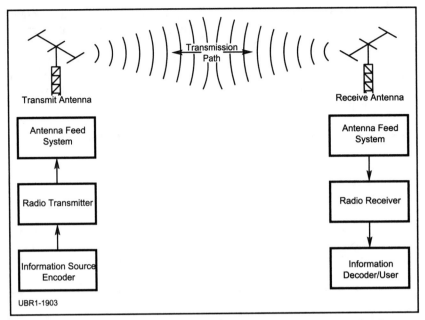

Fig 19-3—Block diagram showing elements of a radio communications system.

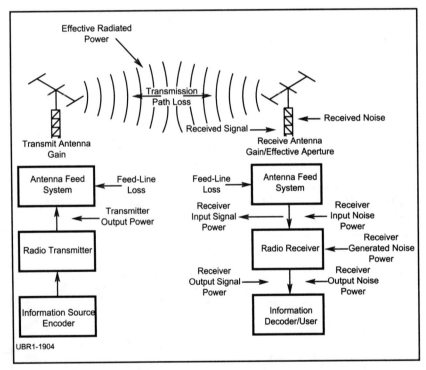

Fig 19-4—Key parameters of a radio communications system.

determined both for antenna systems with an evident aperture, such as a parabolic dish antenna, or a less obvious antenna type such as a thin dipole.

- Feed-line loss—The same type of transmission line loss encountered at the transmitter also happens at the receiver. In the case of the receiver, in addition to the loss of signal, there is an amount of noise generated by the equivalent resistance of the loss.

- Signal and noise power into the receiver—There are a number of noise sources picked up by the receive antenna along with the signal. Different sources dominate at different frequencies, but they all contribute. The ratio of signal power to noise power into the receiver consists of the received signal power divided by the received noise power. It is referred to as the *input* SNR.

- Signal and noise power out of the receiver—All stages of the receiver contribute noise to the output as the signal passes through. This noise is in addition to the noise entering the receiver with the signal. The ratio of signal power to noise power leaving the receiver consists of the received signal power times the receiver gain divided by the received noise power times the receiver gain *plus* the internal receiver noise. This is referred to as the *output* SNR and is always lower than the input SNR. At frequencies below the middle of the HF range, for most receivers, the input and output SNRs are almost the same, however as frequencies increase the internal noise generally becomes more significant.

Yet another topic heading with quotes! *Path loss* refers to the reduction in signal strength as a signal travels from one location to another. A real loss, as I see it, is a reduction in signal level due to power being dissipated in a resistance or equivalent lossy medium. That can be part of path loss, as in a signal with a path through the ionosphere, or through a building, or over lossy ground. On the other hand, a big part of what is considered path loss is the signal reduction resulting from the larger surface area of the sphere surrounding the source. This was shown first in general terms in Fig 15-1, and with the detail of power relationships now shown in **Fig 19-5**. This type of loss is often called "spreading loss" because as the signal spreads out as it travels away from the source there is less of it in any receiver's area of reception.

Path loss is often shown in graphical form as a function of frequency. This is a convenient way of expressing it to allow subsequent calculations to be made based on antenna gain (rather than the antenna size), especially when the same antenna is used for both transmission and reception, as is usually the case in two-way radio and radar systems. Just keep in mind that lower frequencies don't have an advantage here if the physical size of the receive antenna (and hence its aperture) remains the same.

In Chapter 17, we noted that the effective aperture of an isotropic radiator, or receptor was $0.86 \lambda^2$ square meters. At a particular frequency, $0.86 \lambda^2$ will equal 1 square meter and at that frequency, the received power $AP/(4\pi R^2)$ to an isotropic antenna will just equal $P/(4\pi R^2)$ W (for R also in meters). Since $\lambda = c/f$, that frequency can be found by solving $0.86 (c/f)^2 = 1$ for f. This magic frequency is about 278 MHz. Any change to a higher frequency will result in a received power reduced by the square of

f/278 MHz, and conversely. In addition, any antenna gain above an isotropic will increase the received power by the amount of the gain. Thus we could restate the received power as a frequency function as follows:

$$\text{Received Power} = P_{ERP} \, G \, \frac{(278/f)^2}{4\pi R^2}$$

(Eq 19-1)

We can define the path factor as the fraction to be applied to the transmitted ERP to determine the received signal power by dividing both sides by the transmitted P_{ERP} and leaving the antenna gain to be treated as a separate factor:

$$\text{Path Factor} = \frac{(278/f)^2}{4\pi R^2}$$

(Eq 19-2)

where f is the frequency in MHz and R is the range in meters. Note that the power units are no longer specified, if you use watts for P_{ERP}, you get watts received. If you use

milliwatts, you get milliwatts.

To determine the value in decibels, we compute

$$\text{Path factor (dB)} = 10 \log \frac{(278/f)^2}{4\pi R^2}$$

(Eq 19-3)

Remembering the rules for logarithms and squared values, we get:

Path factor (dB) = 20 log 278 – 20 log f – (10 log 4π + 20 log R).

Solving for constants, we have 48.9 – 20 log f –10.99 – 20 log R, which is
Path factor (dB) =
    37.89 – 20 log f – 20 log R

(Eq 19-4)

To find path loss, we change the signs to get a term to *subtract* from the transmitted power:

Path loss (dB) = –37.89 + 20 log f + 20 log R

(Eq 19-5)

with f in MHz and R in meters.

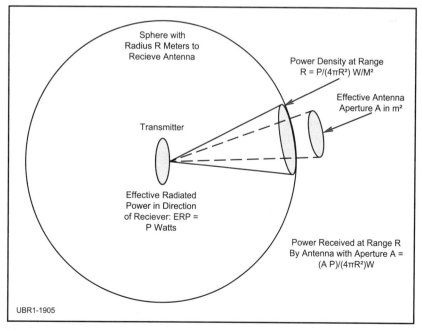

Fig 19-5—Geometric reduction in received signal resulting from receiver being placed at a distance R from the source.

It's possible to perform the system calculations using the various factors and conversions between units as you go, but the use of decibels is a real natural here. If you haven't become comfortable with decibels yet it's worth the trouble to do so before we go further. This is because all the gains and losses are generally expressed in dB and it becomes just a matter of adding them up. It also avoids working with very large or very small numbers, a common source of mistakes.

An Example—a 1-GHz Microwave Link

Let's go through an example of a LOS (line-of-sight) 1-GHz microwave link connecting two buildings 20 km apart. The owners of the business were fortunate that their offices were on the ends of a line-of-sight path. This happens more often in some places than others. The link problem can be stated in a number of different ways, but let's assume some specifications, and assume that the receiver's internal noise is what dominates at this frequency:

1. Equipment parameters
   - Receiver internal noise power: –150 dBW
   - Antenna Gain (T and R): 12 dBi
   - Feed line loss: 2 dB
   - Transmitter power: TBD
2. Operational Parameters
   - 20 km path factor @ 1 GHz: 108.1 dB (from Eq 19-4)
   - Required SNR: 30 dB
   - Required fade margin: 20 dB

Assuming the same antenna and feed system at each end, what is the required transmitter power to meet the design requirements? Let's start at the receiver. We will need to design for a SNR of 50 dB to have our required SNR in the presence of a 20 dB fade from migrating birds. That means that the received signal at the bottom end of the transmission line needs to be: 150 dBW + 50 dB = –100 dBW. The signal at the top of the transmission line must be 2 dB stronger or –98 dBW, and the signal at the antenna can be 12 dB weaker because of the antenna gain, or –110 dBW.

Looking at the transmitter end of the link, the ERP of the transmitter station must be 108.1 dB stronger due to the path factor to have the required signal at the receive antenna. That means that the ERP must be –110 + 108.1 or –2 dBW. Since we have the same 12 dB antenna gain and 2 dB feed line loss, the transmitter must put out –16 dBW. Now we can convert back to power and find that the required transmitter power is just 5.5 mW.

In a real system design problem, the parameters would not be all specified so neatly. Some would have to be converted from other units, however, the principles remain the same. In addition, while we have allowed for fading, we have not provided any allowance for equipment aging or degradation. A typical private microwave system might have a transmitter with a power output of 1 W. The problem can be recast into, "how much extra margin do I have if I buy that kind of equipment?" or "do I need to pay extra for a fancy, high-gain antenna?"

More About Noise

Our example above assumed a specific power level for internal noise in the receiver. While noise can easily be expressed in that form, it will have a value only for a specified bandwidth. It is important to understand this relationship because it is one of the fundamental aspects of system design.

Whether the dominant noise source is internal or external, most noise sources are distributed across a wide range of the frequency spectrum. While each type of noise changes level and relative importance depending on the frequency range (as discussed previously in Chapter 11) within the bandwidth of a particular receiver it can be considered relatively constant with change in frequency. What this means is that the wider the bandwidth of a receiver, the higher the received noise power. Noise level can be expressed in such terms as microvolts per Hz or picowatts per Hz of receiver bandwidth. By multiplying this value times the receiver's *noise bandwidth* (usually the 6 dB bandwidth), the total noise power can be determined.

In this design example, we could have expressed the internal noise level in terms of pW/Hz, and knowing the bandwidth of the receiver (hopefully equal to that required to carry the signal information), we could solve the problem. The number I picked in the earlier example might correspond to the noise power in a not-too-fancy receiver designed for a 48-kHz FDM multiplex group as described in Chapter 10. If we wanted to upgrade to carry a digital 1.5 Mbps data stream (a common digital multiplexing standard from your phone company, designed for 24 channel operation and called a *T1*), to carry both voice and data instead of the 12 analog channels in the FDM group, we might need a bandwidth of 4 MHz instead of 48 kHz.

Since the bandwidth would increase by 4000/48, or by a factor of 83.3, the noise power would increase by the same amount, or a bit less than 20 dB. We would now need to increase our transmit power requirements from 5.5 mW to 458 mW. Our margin for birds nesting in the dish has been reduced significantly. Note that for this to happen, we would have to change a filter in the receiver to establish the bandwidth. We get no advantage with our narrow band system if we buy "upgradeable" radios with the 4 MHz filter already in place and just use 48 kHz of bandwidth. The noise we receive is dependent on the receiver design, not the bandwidth we choose to use. By matching the receiver bandwidth to the spectrum of the transmitted signal we will optimize the SNR, all other things staying the same.

Review Questions

1. What is the minimum clearance of a 30-mile 1-GHz link with the only obstruction 1 mile from terminal one? (Hint, watch your units!)

2. Compute the maximum LOS path length for two stations with 100-meter high towers over a seawater path (at high tide). Assume an operating frequency of 1 GHz. (Hint 1: solve first with a guess at required clearance then redo based on the distances found. Hint 2: the distance in Fig 19-1 is for one tower to the horizon, the other distance to the horizon from the other side is the same.)

3. Redo the 1-GHz example if the fade margin needs to be 40 dB.

Chapter 20

Communications Systems —Going the Extra Mile

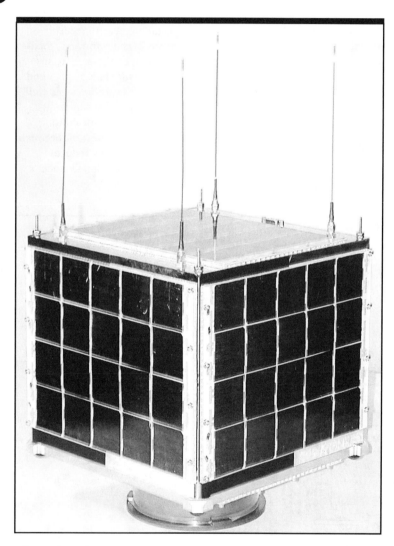

AMSAT-NA microsat AO-51 "Project Echo" satellite, photographed before being placed in orbit in 2004.

Contents

A line-of-sight radio link is limited in distance by the geometry of the configuration. Each fixed link must be analyzed to determine if natural or man-made obstructions in the path will limit communications. An assessment of the sort described in the last chapter is used to establish the radio parameters needed to determine if the path is useable with reasonably tall towers. Then the analysis is completed to determine the ERP required to provide the desired signal to noise ratio. Unfortunately, it is often necessary to communicate further than the LOS path distance and other techniques must be used to enable communication over that longer range.

Point-to-Point Relay Systems

The most straightforward way to lengthen the path is to interconnect multiple point-to-point LOS links to make a system that extends between the desired end-points. **Fig 20-1** shows such a system with one intermediate location. In this case cities A and C have a need for telephone connections between them, but don't have a LOS path that will reach all the way. Because they each have a LOS path to the same mountain top between them, a relay station can be installed there to receive signals from A and retransmit them to city C, and vice versa. Note that while the *terminal* stations at A and C need multiplex equipment as well as radio gear, the *relay* station just needs to have two sets of radio equipment and a connection between them at baseband. This reduces capital expense and maintenance complexity and cost.

In many cases A, B and C are all cities with telephone users. In such a case, the configuration is modified to that shown in **Fig 20-2**. This is called an *add-drop* relay and adds a pair of multiplexers, one on each side of the station. This configuration adds the capability to provide telephone lines between cities A and B and between cities B and C in addition to those going between A and C. All three stations are now terminal stations. Until fiber optic links became dominant in the 1990s, the US was connected by multiple strings of microwave links to provide telephone service from coast to coast. Major cities would have at least two such links to the next cities, typically 20 to 30 miles distant in opposite directions, with some major hubs having both north-south and east-west connectivity, tying together the many

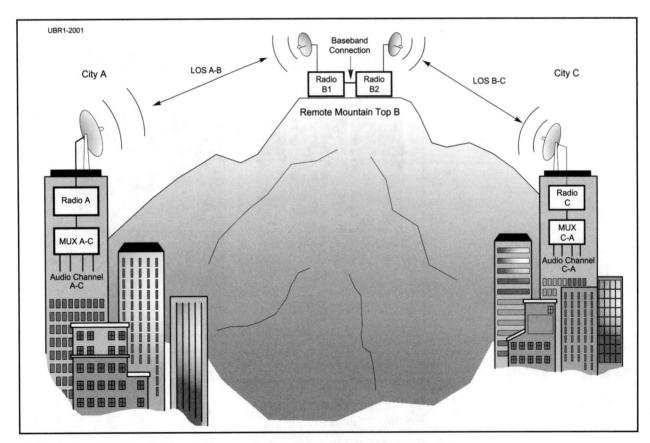

Fig 20-1—Two LOS fixed links interconnected through a baseband relay station.

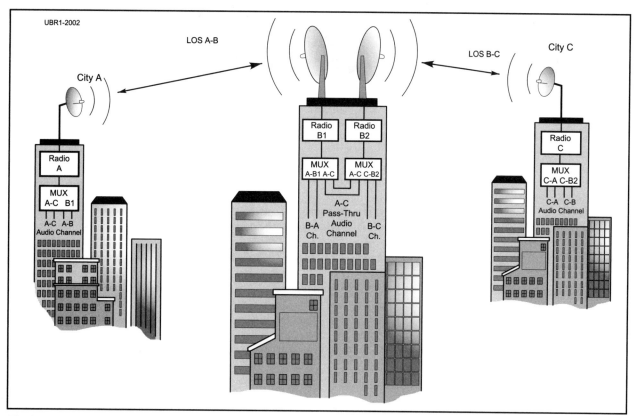

Fig 20-2—Two LOS fixed links interconnected through a drop-insert relay station.

paths that covered the country.

In addition to AT&T, the predominant US long distance carrier, at least, until deregulation in 1984, other carriers such as MCI (*Microwave Communication Incorporated*) had large domestic microwave networks. Fiber has largely replaced such microwave radio connections in the US, although in areas where terrain is hostile, it still is an attractive choice. The military and emergency management teams are equipped to roll in truck-mounted shelters containing microwave, multiplex and portable antennas to establish multi-channel voice and data system connectivity in short order anywhere in the world when they are needed. Your author spent a few years, many years ago, as a soldier serving as a shift leader at US Army microwave stations in Germany.

What Happens to Performance in a Relayed System?

Good question! Unfortunately, it has a multipart answer depending on the type of signaling employed. A system of the type shown in Fig 20-1 has a certain *output* SNR as the baseband signal leaves the first receiver. If it is an analog multiplex system, that SNR becomes the *input* SNR to the next link, which adds its own noise. As a consequence, if both paths have similar SNR, the output of the second link will have twice the noise power, or a SNR degraded by 3 dB. The noise of each link adds as such a signal goes across the country, and thus to achieve a defined SNR, the SNR of each individual link needs to be much higher than the final desired result, so that the combined SNR will meet that specification. In the days in which this was the most common type of long-haul telephone system, a listener could immediately tell if she were receiving a local or long-distance call based on the background noise level.

Today it's different, largely because most signaling and multiplexing is performed by digital, rather than analog, systems. With a digitally encoded voice system, there is an equivalent "noise" introduced into the system as each voice sample is encoded into a digital word. This is due to the fact that, however small the size of the least significant bit, the sample must be encoded exactly as a 0 or 1, but is actually somewhere in between. This is similar to "rounding error" in decimal systems. The difference between the actual value and the value of the number representing it acts just like noise and has the name—*quantizing error* or *quantizing noise*.

A benefit of this type of digital encoding is that as long as a received signal is sufficiently above the noise to allow a determination of whether each bit is a zero or a one, the only noise will be the quantizing noise introduced in the encoding process at the first terminal. This continues through any number of relays, as long as we can detect the correct bit value and regenerate a new pulse train. If we decode the audio and then re-encode it, we will introduce additional quantizing error at each step.

A system with a similar looking diagram, but with different sets of applications, is a repeater used to extend the reach of vehicle-to-vehicle communications systems. Such systems, whether public service, utility vehicles or Amateur Radio operations, tend to be single-channel, push-to-talk (PTT), half-duplex, FM radio systems operated in the various VHF and UHF bands allocated to their service.

A key difference between a *repeater system* and the previous relay network is that the latter needs to support wideband multi-channel communications in both directions simultaneously. On the other hand, the repeater receives signals from either, but not both, vehicle stations and retransmits to the whole region on a different frequency, as shown in **Fig 20-3**. If both stations choose to transmit (or receive) at the same time, neither will hear the other.

Another difference is that while the relay system needs to communicate with a particular fixed destination and can take advantage of an antenna beamed in that direction, the repeater must receive from both sides of the obstruction and transmit to both sides also. Note that while the mobile stations can use push-to-talk (PTT) to change between receiving and transmitting, the repeater must simultaneously receive and transmit (on different frequencies, of course) to achieve its function. This is generally accomplished by having F1 and F2 far enough apart so that the repeater transmitter does not overload the repeater receiver. This is accomplished by using appropriate filtering. Sometimes the same antenna is used for both transmit and receive, requiring particularly effective filtering to keep the, say, 100-W transmitter signal from interfering with the sub-microvolt receiver!

In addition to the antenna connections, the audio from the output of the repeater receiver is fed to the microphone input of the repeater transmitter. Most repeaters are designed not to transmit unless the receiver is receiving an input signal. A *carrier operated relay* (COR) monitors the receiver AGC voltage and turns on the transmitter when a signal comes in that is strong enough to relay.

In some systems special tones are used to restrict repeater operation only to stations sending the special tone signal, which is typically much lower in frequency than voice communications audio.

Trunked Systems

A kind of combined application of these concepts is to use point-to-point links to extend the range of mobile stations. Picture a microwave LOS link located on the same mountaintop as the repeater station, just discussed. When the repeater receiver receives a signal, it not only turns on the local repeater transmitter, but also passes an audio signal down one channel of the microwave multiplexer to the next mountain. Along with the audio is a signal indicating that the COR line of the repeater has been activated. At the next mountain, the microwave receiver's remote keyed channel is tied to a repeater at that mountain and operates the COR of that repeater. Thus, a car within range of

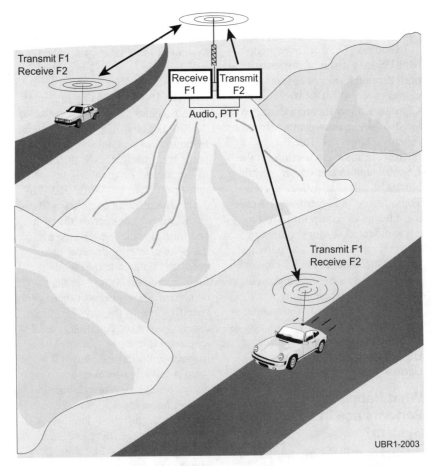

Fig 20-3—Repeater system to extend car-to-car range.

Transmit F1
Receive F2

Receive F1

Transmit F2

Audio, PTT

Transmit F1
Receive F2

UBR1-2003

the repeater at the first mountain can communicate with other cars and offices there, or equally as well with vehicles or offices at the second. Depending on terrain, the stations can be located so that the vehicle always has contact through one repeater or the next, perhaps the whole distance of an interstate highway.

The microwave system can support multiple mobile channels by having a multiplex channel assigned to each, For example, there may be one or two channels assigned to different repeater stations for the state police, another one or two for the highway department radios, etc. Additional channels can be used to support communications services between fixed locations such as state offices along the route. Next time you drive down a major highway, note the microwave towers often located near state police barracks or maintenance depots.

Another category included in the group of repeater communications systems is the now ubiquitous cellular telephone system. Cellular networks have a lot in common with the remote-keyed repeaters of the previous section, except that cellular systems are provided by telephone companies and are shared among large numbers of users through the reuse of channels in nearby areas.

Cellular systems were developed to replace the Mobile Telephone System offered by the telephone companies. These used a limited number of VHF channels. Each channel would cover a substantial range, typically a metropolitan area, served by a high-powered base station with a high antenna. The VHF system worked reasonably well, however, the limited number of channels and wide coverage made it impossible to meet demand in most major markets. The cellular telephone system took a different approach by using low powered transmitters, low antennas and the ability to reuse channels across an area.

The modern cellular telephone has amazing capability built into a very small enclosure. We will concentrate on the radio side of cell phones here, and leave the cameras, video, text messaging and hi-fi subsystems to others. There is plenty to talk about for just the radio!

Analog Cellular Systems

In the beginning there was analog. The first US cellular systems made use of a range of frequencies removed from the upper channels of the UHF TV band. These ranged from 825 to 894 MHz. They were set up in two bands of 832 30-kHz wide channels, one going from the cell phone to the cell site, the other from the cell site to the user. See **Fig 20-4**. Each cell tower is designed to serve a relatively small area. Adjacent cells all operate on different channels so

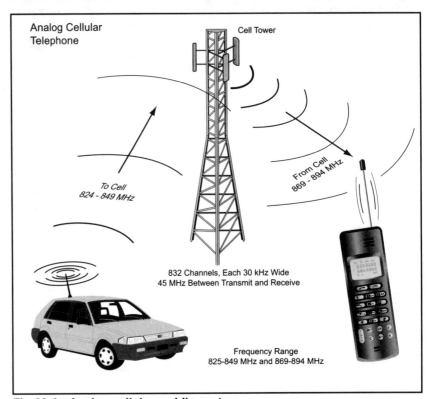

Fig 20-4—Analog cellular mobile system.

there is no interference at cell boundaries. However, the next cell out can reuse the same frequencies as the first cell, as shown in **Fig 20-5**. This allows much more efficient use of the radio spectrum. Of course, this is not without some cost. Many more base systems and towers are required, as well as control logic to keep track of each mobile station's location while they're using the cell system.

In operation, a cell phone starts out listening and transmitting on one of the 42 shared service channels, typically 6 per cell. The cell phone transmits a notification signal periodically that is received by all cell towers within range, typically only a few at one time. The cell towers communicate among themselves, often by a wire link, to decide which cell has the best reception. The tower that has the best reception

initiates any calls to the mobile phone. The mobile phone continues to monitor the service channel and is notified of an incoming call. If it responds, the mobile phone is assigned one of the typically 56 channels assigned to that cell and the call is completed there. This approach is called *frequency division multiple access* or FDMA. This means that multiple users can be served simultaneously by assigning them to different frequency channels.

In an effort to minimize interference, power management is used. If the cell receives a strong signal from the mobile, the mobile is directed to reduce power. As the mobile unit approaches a cell boundary, the adjacent cell notes that the signal is increasing, while the first cell notes a decrease. As the boundary is reached, based on signal strength, the call is *handed off* to the adjacent cell.

The mobile is assigned a new operating frequency by the cell tower in the second zone and the mobile unit switches to one of the 56 channels assigned to that zone.

Simultaneously, the cell switching equipment changes from routing the call via the first cell tower to the second. The amazing thing is that this all works most of the time and the user is completely unaware that it is happening! Obviously, the cellular system is made possible by having compact, high-speed processing hardware in the cell phone. There is really more "processing" than "radioing" going on in such systems.

Digital Cellular Systems

The analog cellular systems were very successful, and they worked well wherever there were cell towers to support them. There were a few limitations, however. A big concern was privacy—regular telephone users had come to expect a certain freedom from eavesdroppers, and there was essentially none provided in an analog cell system. Anyone with a scanner receiver, or an old television set that picked up all the original UHF channels using an analog tuner, could (and still can) listen to cell calls in their neighborhood. US regulatory bodies made rules prohibiting scanners that covered the cellular frequencies from being imported or sold, but there were a lot of them out there.

Another limitation was found when trying to use analog cellular phones to carry data using modems. The data rates were not very high and channel switching between service and operating channels introduced errors.

The use of digital encoding in cell systems solved these problems, at least to a certain extent. The encoded voice was still not encrypted, so anyone with the appropriate equipment could still eavesdrop. However, at least the casual listener could not overhear personal conversations.

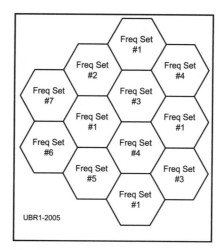

Fig 20-5—Analog cellular system—illustration of possible frequency reuse scheme.

Digital data could be sent with the higher throughput of the digitally encoded channel once appropriate interface standards were established. An added benefit is that three times as many channels can be established with efficient coding.

The first digital systems used *time division multiple access* (TDMA), dividing up a bit stream that used the full bandwidth of the channels assigned into time slots sufficient to carry a coded sample of voice information. As in the case of FDMA, a number of service time slots are reserved for control purposes, including the ability to direct a mobile phone to switch to a particular time slot in order to complete a call.

Initially, digital systems shared the same 800-MHz frequency range channels that were used by analog systems. Later a range of frequencies around 1900 MHz was allocated for digital mobile telephone service.

Another access technology used for digital mobile service is called *code division multiple access* or CDMA. In this arrangement, each station transmits a pulse train at the full data capacity of the assigned band. This will typically be more

than 100 times the data rate required to support the encoded voice traffic, so each transmitted bit is converted into a, let's say, a 100-bit code word. Each station is assigned a code word, much in the way that they are assigned a frequency in FDMA, or a time slot in TDMA. At the receiving end of the channel, the receiver multiplies the incoming data stream by the bits of the code. At the end of the code word the results are added and if the same code word was sent the result should be 100 times as strong as any of the received bits.

To send a value of a binary "one" the code word is sent and 100 is decoded. To send a "zero" each bit is inverted and a −100 is decoded. A station assigned a different code word will decode a much lower value, hopefully zero instead of 100, if the codes are properly selected. Multiple conversations can occur simultaneously using a set of appropriately related code words. There have been many discussions about whether CDMA or TDMA is the better approach, and proponents of each can construct examples where one or the other has the advantage.

The above systems are all used in the US. The rest of the world has adopted the *Global System for Mobile* (communications) or GSM. GSM is a digital service that uses a TDMA access method similar to that used in the US but different in specifics. GSM service is available in more than 100 countries, including most of Europe and Asia. It is possible to purchase cellular phones that support both US and GSM offering, essentially multiple phones in the same case. Service usually must arranged for each separately, often with multiple providers. However, dual US/GSM capability is a major convenience for those who engage in international travel.

The space age began in 1957 with the surprise launch of the USSR (now CIS) experimental satellite Sputnik I. This satellite transmitted a signal at 20.05 MHz that could be monitored worldwide as it went around and around the Earth. Years before, a number of scientists recognized the potential of having not just a transmitter, but a wideband relay in space that could allow the transmission of first voice, and then other types of signals between distant spots on the Earth. Early experiments with passive satellites, such as metalized balloons, provided interesting results but did not turn into commercial successes.

Geostationary Orbit Satellites

While the launch rockets to get there were yet to come, for many years those involved with the study of celestial mechanics recognized that for an object to orbit the earth, it would need to attain a certain velocity. This would balance flying off into outer space and falling back into the earth's atmosphere because of the pull of gravity. At low altitudes, the velocity would result in the satellite appearing to move rapidly past objects on the Earth's surface. At very high altitudes the Earth would appear to move past the satellite. There is a special altitude— 22,300 miles above the Earth's equator—at which a satellite will appear to stand still above a point on the earth. This kind of orbit is known as a *geosynchronous* or *geostationary* orbit and is perfect for some communications satellite purposes. As shown in **Fig 20-6**, a satellite in such an orbit is far enough away so that it can see almost half the Earth's surface and thus can relay signals between any two points within its coverage shadow.

Although space is a pretty big place, there is only room for a finite number of such satellites so that they can operate without interfering with either each other or with each other's

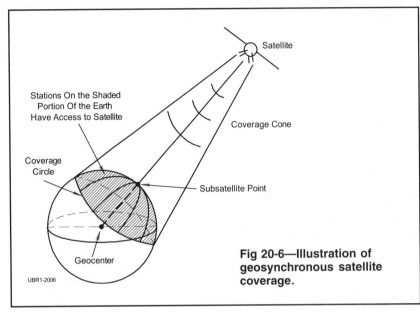

Fig 20-6—Illustration of geosynchronous satellite coverage.

ground stations. By international agreement the slice of space at 22,300 miles above the equator is divided into *orbital slots* 1.5° apart, with reuse in different frequency bands. That allows for exactly 240 spaces. Of course, not all orbital slots are equally popular—there may be less demand for slots over the Pacific than over the Americas or Europe or the Middle East, for example. The width of each space is around 700 miles; however, satellite operators are required to station keep, which means that their satellites are constrained to a much smaller volume of space, typically 50 miles square by 75 miles in altitude. This allows receive antennas to focus on one satellite without receiving interference from its neighbors. Such flight operations, stationkeeping as well as antenna orientation, require fuel for small thruster engines and when the fuel is gone the satellite will eventually drift off into space. It's been some years since I've been in the satellite business, but I remember that a satellite cost about $50,000,000 to buy, about the same

to launch, and almost as much for launch insurance. It's good to plan the satellite driving carefully!

Space Assets

Satellites function a lot like a mountaintop repeater, but there are some significant differences. Like the repeater, the satellite receives signals on one frequency and retransmits on a different frequency. Unlike the repeater, which does this for a single channel, the satellite *transponder* receives a whole range of frequencies, often consisting of multiple signals. A typical transponder might have a bandwidth between 36 MHz (enough for 6 standard television channels) up to 64 MHz. Usually there will be at least enough transponders to fill the 500 MHz of bandwidth of frequency allocation. Most early satellites operated in C-band[1] with an uplink of 5.9 to 6.4 GHz and a downlink of 3.7 to 4.2 GHz.

Later satellites, starting in the early 1980s when equipment technology caught up, also operated on the K_u band, usually with an uplink of 14 to 14.5 GHz and a downlink of 11.7 to 12.2 GHz. K_u band has the unfortunate property that rainfall results in significant losses, since the resonance of water molecules is in the same band. One satellite system I worked on had a K_u band station in central Florida, where it rains every afternoon, and that station required a much higher gain antenna.

A typical satellite includes multiple antennas. Antennas can split power between the transponders, or can be used separately, providing beams to particular areas. For example, a satellite over the Pacific might have a wide-area antenna covering the whole region, with another antenna focused on Hawaii or New Zealand to provide improved performance in those regions. Some transponders can be dedicated to each type of service, or they can be changed by control station command. A user can thus contract for service beamed to a particular region.

Each satellite transponder is usually a wideband analog system. Transponders designed to support analog service can also be used to process digital signals. To support analog systems, perhaps with six TV signals in a 36-MHz transponder, the total power that a transponder must handle must be below the maximum saturated power that the transponder can transmit, perhaps with 6 dB of additional "overhead." This will avoid intermodulation products interfering with other signals in the band.

On the other hand, if the transponder is used to support only a single high-speed digital pulse train, intermodulation is not an issue and the transponder can be driven right up to saturation without causing any problems. This provides a significant SNR advantage to digital signaling through satellites. The satellite provider doesn't need to decide. If a user wants to contract for high-speed digital use, the same transponder will work just fine, and it will provide that 6-dB benefit without having to do anything special.

So What's the Problem?

It sounds like satellites can provide great service for all kinds of applications, so why don't we use them for everything? Good question! Satellites have two major problems for use as transport for voice service. First there's delay. It takes almost a quarter second for a signal to get up to a geosynchronous satellite and back down. Now if you're talking to someone and ask a question, that means it takes almost half a second to get the beginning of an answer. Some answers are worth waiting for, but most people would assume that the other end was struggling with the answer and start to ask the question a different way, just as the answer starts to come in. It can be very difficult for those used to rapid-fire communication!

The other problem has to do with cost. Terrestrial communication tends to have a distance component to cost. Satellite communication has the same cost wherever the end points are, if they're in the same footprint. There is some break point at which satellite communication is cheaper than terrestrial communication. Back when I was in the business, the break point was at 400 miles. If the two end points were more than 400 miles apart we could offer a less expensive service.

Well, the cost of terrestrial service—based on the explosion of fiber based capacity—kept getting cheaper. Satellite service still costs the same; after all, we had already launched our satellites! The break-even point moved farther and farther away until there was no place in the US in which satellite service costs less. Besides, fiber was quiet, reliable and had no delay. You don't see a lot of voice satellite services offered in the US these days.

On the other hand, satellites are very efficient in a situation in which a single signal must be delivered to multiple destinations—satellite TV is just such a situation. With terrestrial services, typically there must be a link to each destination. With satellite, it costs the same whether that TV signal is sent to one destina-

tion or to 1000 cable operators across the country. That's why most TV distribution is via satellite.

There are many other services that also can be efficiently carried by satellite—for example, stock ticker information and other kinds of data distribution. Voice service is still viable in special cases, such as with ships at sea or other spots where terrestrial service is not available.

Low-Earth Orbit Satellites

While geosynchronous satellites are the 800-pound gorillas of satellite services, there are other ways to make satellites work. *Low Earth Orbit* satellites or LEOs have a number of advantages. Unlike geostationary satellites, LEOs go whipping by overhead quite rapidly. For many kinds of services, this requires ground stations that can predict when a particular satellite will be in range, steering the ground station antenna towards it and then shifting quickly to the next satellite when the first one leaves. Such LEO satellites have much smaller footprints. See **Fig 20-7**, for example. A smaller footprint means that additional interconnectivity is usually needed—either via terrestrial links or from satellite to satellite—if long distance coverage is required. On the other hand, the problem of delay is significantly reduced because the LEOs are closer to the ground stations they service, and much less power is required at both satellite and ground station ends of the link.

An interesting joint venture resulted in the design and deployment of a LEO system intended to provide voice service where there were no terrestrial facilities available. The *Iridium* system was designed to provide voice services between users of small hand-held cell-phone like terminals and other Iridium users, or with the rest of the world's wired network. It did so with no on the ground investment required, but the usage costs were very high to pay for the expense of the many satellites required to deliver the service. Iridium was not a commercial success, but it was such a useful system for government users in remote parts of the world that it is still in service under different ownership and will likely continue in operation for some time.

Other users also make use of non-geostationary satellites. We will discuss the *Global Positioning System* (GPS), a non-geostationary system, in a later chapter. Amateur Radio operators have had a succession of satellites available for their use for many years. Typically constructed by amateurs and launched in "space available" compartments of launch vehicles, these have been put into orbit by amateurs from many countries since the early 1960s.

Notes

[1]Letter designators were assigned to microwave bands in the early days of radar to aid in security. They are still in use today, both in radar and communications terminology.

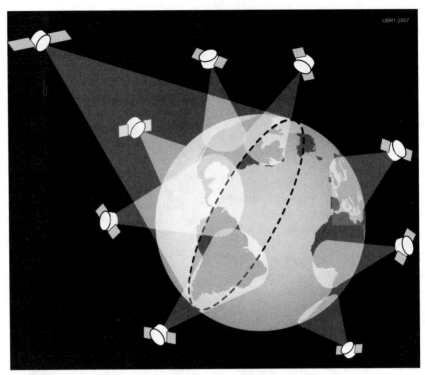

Fig 20-7—Illustration of low-earth-orbit satellite system coverage, compared with geosynchronous satellite coverage.

Review Questions

1. Why do you think fiber has displaced radio communication in the US? Under what circumstances might radio still be a viable alternative?

2. Repeater systems tend to be PTT (push-to-talk) from a vehicle's perspective. What would be required to convert such a system into a full duplex system, mobile station-to-mobile station, so that someone can interrupt in mid-sentence?

3. Consider a trunked radio system running along a highway. Why might there be a need to split the highway into multiple sectors?

Chapter 21

Ground-Wave and Ionospheric Communications

Talk about a direct line-of-sight path: EI2AIR balloon mobile rises towards the sky!

Contents

Fortunately, radio communication is possible beyond purely line-of-sight (LOS) distances. Some methods to provide communication beyond these distances were discussed in Chapter 20. They either made use of multiple line-of-sight "hops" or via LOS-reachable repeaters, on the Earth or out in space. Such systems are really a special case of LOS, although clearly they are effective for long-distance applications. A disadvantage of those systems is that you need to control some real estate (or space) in between. In both cases, significant pre-arrangements are required. In this chapter we will be talking about getting beyond the LOS threshold in radio hops, with only nature in between.

Ground Wave Propagation

A commonplace communication method on LF and MF is called *ground wave communication*. This works much like LOS communication, except that it continues well past the visual horizon. The signal generally is launched by a vertically polarized antenna, which has significant energy focused towards the horizon. This is the same place we want it for LOS use: tangent to the Earth's surface. As the ground wave propagates, interaction with the ground results in a slowing of the propagation velocity in the region near the Earth's surface. The resulting wave takes on a tilt, with

the upper regions going faster than the lower. As a consequence the wave front continues along the local surface instead of shooting off towards space. A comparison between LOS and ground wave propagation is shown in **Fig 21-1**.

Ground wave efficiency is very dependent on ground characteristics. If the conductivity is low, the resulting lossy earth quickly dissipates the power at the surface and less is left to propagate onward. In areas of higher conductivity, and especially areas without interfering terrain features, propagation via ground wave can extend out more than a hundred miles.

While ground wave propagation works day or night, the LF and MF frequencies used are subject to nighttime ionospheric skip propagation, as will be discussed later in this chapter. A signal arriving from outside the region may be much stronger than the ground wave signal and this can result in interference. In the US, certain AM broadcast stations are designated as *clear channel* stations and other stations sharing the frequency during the daylight hours are required to go off the air at dusk to avoid interfering with the clear channel stations at night.

Ionospheric Communications

In Chapter 15, I described the ionosphere, its virtual layers and their effect on radio communications.

While the concepts are straightforward, the application of them is anything but! In Chapter 15, I made note of the various frequency-dependent operational parameters of the ionosphere — the LUF (lowest usable frequency), MUF (maximum usable frequency), FOT (frequency of optimum transmission) and the critical frequency. Optimum use of the ionosphere requires the knowledge about all these frequencies, as well as knowing whether the end points are in daylight or darkness.

The frequencies that describe ionospheric properties change throughout the day, and they can be very different from one day to the next. Ionospheric properties respond to solar activity that occurs over multi-year cycles, and they are at the same time influenced by dynamic daily changes on the sun.

Once you know the frequencies, it's pretty straightforward to predict how to propagate a signal from one point to another. The resulting signal levels can be predicted based on a variant of the analysis technique we described for LOS propagation. The path loss will be at least as great as the LOS path loss over the total path distance, but generally a lot more, since only a fraction of the signal energy gets refracted through the ionosphere. Some paths follow a multi-hop route, and each encounter with either earth or ionosphere results in additional loss. So what are we to do to make path predictions?

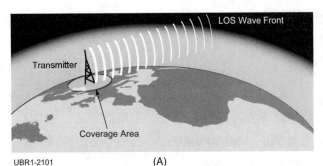

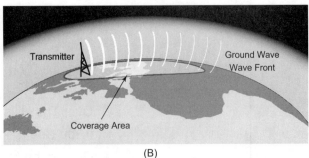

UBR1-2101   (A)   (B)

Fig 21-1—A line-of-sight signal at A, compared to a ground-wave signal at B.

If You Can't Figure It Out—Measure It!

In this chapter we will generally be talking about long-haul HF, often on different continents. One way to solve the problem about which frequencies to use for ionospheric communication is to use a different communication system to coordinate trial frequencies and check out the performance of different HF frequency ranges. This is a variation of the cellular telephone TV ad, "Do you hear me now?"

While that might be effective, it begs the question, if you have another way of getting there why would you worry about using HF in the first place? A more useful way is to use common data. It is pretty straightforward to determine if both stations are in daylight or darkness, then it's a matter of figuring out what frequencies will support communication at that particular time. Because of the variations in HF performance, each user typically has a number of assigned frequencies throughout the HF spectrum so that different times and paths can be accommodated. The following are techniques that are used to establish communication in such a variable environment.

Trial and Error

Some systems, for example shore stations in the maritime HF service, monitor all their assigned frequencies all the time. If you're at sea, you look up the frequencies of the, let's say, Wilmington, North Carolina shore station, set your radio to one of the frequencies and call—"Wilmington marine operator, this is the sloop *Windfall*, WRE3053, over." If the operator hears you on that frequency they will respond on their associated transmit frequency. If you don't get a response you try another frequency. If the operator does answer and finds you weak, they may inquire about your location and suggest a different frequency that has previously worked into your area. This process can be

systematic, based on prediction software or channel monitoring, or it can be totally random. When you hit one of the Wilmington channels that works, you can start communicating—it doesn't matter how you got to that frequency.

Scheduled Trial and Error

There are likely more HF operations in which operators are not equipped to monitor multiple channels 24 hours a day. For these situations, the operators at each end can select the frequency to use based on a prearranged schedule. At the desired time, station A will call station B on the highest assigned frequency for, let's say, 30 seconds of each of five minutes. If no response, station A will call station B on the next lower frequency for five more minutes, and so on until they find a frequency that does work satisfactorily.

Note that they should probably start at the highest frequencies, since if they work at all, the losses are generally lower on higher frequencies. This method requires only moderate clock synchronization to be successful, although it may take half an hour to establish communications. This may be suitable for some applications, but probably not if you're taking incoming gunfire, such as a military vessel might!

As a variant of the above "brute force" method, each operator can make use of a propagation-prediction program based on available sunspot and solar activity data. Such programs can provide a probabilistic prediction of MUF, LUF and FOT. It is important that both stations have access to the same basic data so that the programs can predict the same frequencies for each. Then the operators just use the information on an agreed upon basis to allow the trial of a smaller set of frequencies. This can reduce the time it takes to

establish communication, sometimes a critical operational factor.

Predictions Based on Measured Data

A somewhat different approach can make use of observed reception data prior to initiation time for the actual communications. In this case, each station can listen to either cooperating stations (beacon transmitters established for the purpose), or non-cooperating transmitters—perhaps well identified short wave broadcast transmitters originating near the far-end station. Each can then draw conclusions about which frequencies might be useful. This is particularly useful if station B has multiple receivers tuned to a number of the likely frequencies and can determine what frequency station A has selected. Otherwise it is quite possible that each station will draw different conclusions and they will never connect up together.

Each station can also be equipped with *ionospheric sounding* equipment. Such equipment is designed to send signals at progressively higher frequencies straight up toward the ionosphere. If the frequency is reflected the equipment will make note of the amplitude of the received signal and the propagation time from transmission to reflection return. The system can determine the LUF and MUF. By looking at the delay times, the actual layers resulting in reflection can be determined for each frequency. This data can then be used as input to a propagation-prediction program that can determine the most likely optimum frequency range to go to the far-end station.

The problem with this system is that both ends may observe different results, particularly in different regions of longitude. It is most effective if one station makes the decision and the other can monitor all the assigned frequencies and then respond on the best one where the first station can be heard.

Automatic Link Establishment

Modern tactical HF radio equipment is generally not equipped with sounding equipment or computational facilities and may be carried on a soldier's back or in a vehicle. Further, the soldier may need to communicate quickly, with minimum operator decision-making. *Automatic link establishment* (ALE) is a technique that automatically tests each assigned frequency in synchronism with the far-end radio and determines the optimum frequency to use at that moment, for that path. This obviously requires special radios with compatible processors and a high degree of clock synchronization, or else the radios will not be on the same frequency at the same time.

Also required are antennas that either cover the whole range of frequencies, or that can tune quickly based on commands from the processor. The radios take turns sending a special coded message including an identification tag. Each radio measures the SNR at each frequency and decides on the optimum frequency for each direction (often, but not always, the same). The operator only needs to know to hit the ALE button and stand back. In seconds, if there's a path, the radios will find it and tune to the appropriate frequency.

In the last section, we discussed finding out what frequencies should be used to get an HF signal to a desired destination via the ionosphere. The next problem is finding out how to make your signal go where you want it to go to communicate with that distant station.

Pointing The Antenna

Not all HF antennas are intentionally directional, but as we noted earlier, an isotropic antenna is hard to actually make. Thus antennas will almost always radiate in some directions better than others, both in azimuth and elevation. In order to optimize antenna direction, you need to know the desired path between the two end points. We will consider two general categories.

Near Vertical Incidence Skywave (NVIS)

For many years, relatively short range tactical military communications have relied on VHF LOS systems. These are generally easy to carry, require small and easily deployable antennas and provide reliable and predictable communica-

tions, so long as a LOS path can be established. US forces supporting operations in the Balkan region found themselves in an area in which the mountainous terrain would not support LOS communications for many desired tasks. The paths were not needed for a sufficient term or with sufficient notice to allow

establishment of intermediate repeater facilities. Military satellite communications is one solution, however, that generally requires rather careful antenna pointing, not always possible in tactical environments.

The solution was found on HF using *near vertical incidence sky*

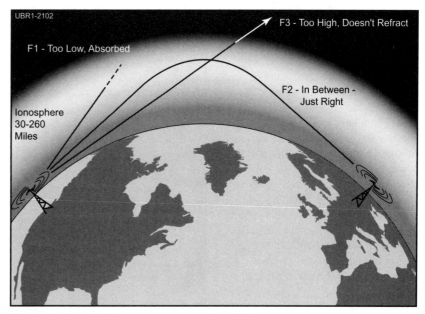

Fig 21-2—Establishing a frequency for HF ionospheric communication.

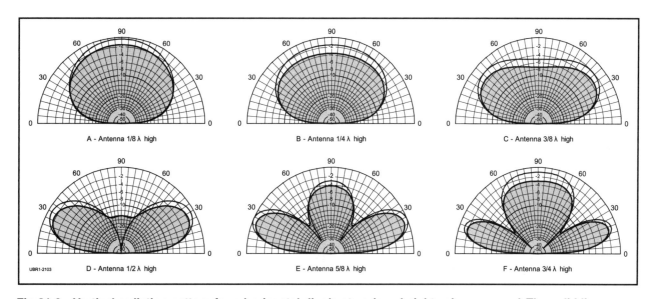

Fig 21-3—Vertical radiation pattern for a horizontal dipole at various heights above ground. The solid lines represent pattern over perfect ground, while the shaded patterns represent typical real ground.

wave or NVIS techniques. In this approach, HF signals are transmitted almost directly overhead, to be reflected downward via the ionosphere to a location perhaps only 25 miles away. The technique can provide reliable communications out to 100 miles or more depending on the elevation angles employed. A nice feature of this methodology is that no special equipment is required. Existing radio systems, originally designed for long-range communications, can be pressed into service. The only modification required was to lower horizontally polarized antennas so that the radiation reflected from the ground added in the upward direction.

Notice in **Fig 21-3** that the height of a dipole, or any horizontally polarized antenna, has a major impact on the vertical radiation angle. At a height of $^1/_8$ λ or less, most of the radiation from a horizontal dipole antenna goes skyward, not towards the horizon. As we will discuss later, this is exactly what we don't want for long distance communication, but for NVIS it's just right. It's almost always easier to install an antenna at a lower height than at a higher height, so this is an easy modification to the usual HF operation.

Long-Range Ionospheric Communication

To make HF communications work over long distances, we need to get our signal pointed towards the right region of the ionosphere that results in radiation returning to the Earth in the desired target area. There are actually multiple places that can work between any two points on the globe, and it is important to understand the implications of this.

A radio signal will generally follow a *great-circle path* between any two points on the Earth. A great circle represents the shortest distance between two points on the globe if you could stretch a string between them. Often people are surprised to discover the direction of the shortest path between two points because

Fig 21-4—Antenna system at W1AW, the headquarters station of the American Radio Relay League in Connecticut. In the foreground, the 120-foot tower has Yagi antennas pointed towards the Southwest and Northwest to provide coverage across the US on all amateur bands from 7 to 30 MHz.

they are used to a Mercator map projection. For example, a Mercator projection makes Rome in southern Europe look like it is due east (90° azimuth) of Hartford, Connecticut. In truth, the great-circle bearing for this path is 58°—not 90°.

In addition to that issue, which doesn't take long to recover from, there are two great circle paths between any two points on earth. For obvious reasons, one is generally referred to as the *short path*, the other the *long path*. This is important

because, while the short path might seem like it would have less path loss, solar conditions may result in a usable route via long path and none on short path. Moral—always watch your back!

Many HF station have directional antennas that focus their radiation towards the desired target. In some cases they are fixed antennas, perhaps multiple dipoles phased together and in front of a screen reflector. This kind of *curtain array* (because it looks rather like a curtain) is popular among short-wave broadcast stations. In other installations a station might use rotatable arrays, able to point in a particular direction as their traffic and destinations dictate. The Yagi antenna arrays at W1AW, the headquarters station of the American Radio Relay League, are shown in **Fig 21-4**. These antennas are designed to deliver signals on many frequency ranges to receivers across the US. Other HF users, such as International broadcast stations with requirements for long distance communications, may use similar Yagi systems.

Fig 21-5 shows the transmitter equipment at W1AW. Each rack holds a transmitter and a 1500 W power amplifier used on a single amateur frequency band. The multiple racks are generally used together for simultaneous broadcast on all the major MF and HF amateur bands. At the left end of the complex is an antenna patch panel to maximize connection of any of the transmitters to any of the antennas to allow operational flexibility. Commercial shortwave broadcast stations might have similar layouts, although they are likely to transmit at higher powers, 10 kW to 1 MW, to maximize worldwide reception by listeners with rudimentary, low performance receiving equipment.

Fig 21-5—Transmitter equipment at W1AW. Each rack contains a 1500 W CW and SSB transmitter to broadcast on all amateur bands from 1.8 to 30 MHz for daily information bulletins and code practice.

Propagation by Scatter

There are other mechanisms that can sometimes provide communication beyond LOS distances. One general category is called *scatter*. This refers to waves entering a region above the Earth where some sort of phenomena can cause random reflections of the waves back to Earth. Different mechanisms can come into play, including the ionized trails of meteors entering the Earth's atmosphere, or ionized areas created by high-energy particles ejected by the sun (that can also result in aurora), or even an aircraft within the view of each end point. The height of the scattering region above the Earth determines the possible range of communication, since both ends of the path must be able to "see" the common region.

You can easily imagine a scattering mechanism in which two stations are located on the opposite sides of the flight path into a busy airport. If there is sufficient traffic so that there are always aircraft in view of both stations, some transmitted signal (for wavelengths smaller than the aircraft surface) will be reflected from the underside of each plane. A portion of the reflected energy will reach the distant receiving station. If the aircraft is an altitude of 2000 feet, communication over distances of about 200 miles is possible.

Note that significantly more power is required for scatter communications than would be predicted for a direct LOS path for the following reasons:
• Only a small fraction of the power reaching the scattering surface is scattered in the desired direction.
• Because the scattering surface acts like an antenna reradiating the signal, the signal level will be reduced by the square of the distance to the receiver. The power reaching the scattering surface will also have been reduced by the square of the distance from the transmitter. Thus the re-radiated power will be reduced by the fourth power of the distance. This results in a signal that drops off quite rapidly with distance.
• In the case of our aircraft example, there will likely be a number of scattering surfaces in common view and each results in signals that combine at the receiver. Since both scattering objects are moving, the changing phase will sometimes result in periods during which the signals tend to add together. Stronger signals are good. But scattered signals can also cancel, which can result in severe fading. Fading is bad.

Tropospheric Scatter

The *troposphere* is a region situated below the D-layer of the ionosphere. The transition of radio waves from an electrically neutral region like the troposphere to an ionized one like the D-layer acts a lot like a conductive surface. This results in refraction and scattering of VHF signals. *Tropospheric scatter* can commonly support medium-range communication— beyond LOS. The geometry is shown in **Fig 21-6**. As with all scatter modes, high power is required to get signals reliably to the distant end; however this is a usable mechanism under most circumstances and is commonly employed as a portable military system for transmission up to around 800 miles.

Meteor-Trail Scatter

Working our way up to progressively higher altitudes, we next come to *meteor trails*. As the Earth races around the sun and together they pass throughout the Milky Way, the Earth's atmosphere is constantly encountering dust and fragments from meteors, comets and other space debris. In most cases, the material is burned up in the atmosphere and never encounters the surface (I'm glad to say!). The process of burning up results in a plasma like region in the atmosphere that is sometimes visible. Even though visible activity is infrequent, there is almost always some of this happening and the plasma regions can scatter radio waves in the VHF region. Because of the wide range in scattering density, large amounts of power are needed. The rapidly changing phase shift also results in a restricted bandwidth, typically about one or two voice channels.

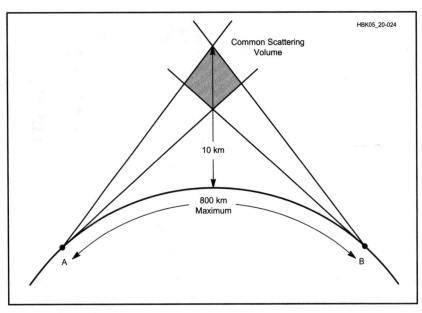

Fig 21-6—Geometry of VHF tropospheric scatter communication.

Review Questions

1. Under what conditions would NVIS HF communications be preferable to VHF LOS communications? When would VHF be better?

2. You are the Circuit Chief of a HF radio circuit operating between North America and Europe. You have been assigned frequencies of 4, 8, 12 and 18 MHz. Discuss which might be preferred, under various times and solar conditions.

3. You set up a ground wave path at 2.1 MHz between two stations 100 miles apart during the day. At night you find that you are not hearing the station. Instead, you hear someone 1200 miles away. What do you think is happening and what should you have considered while the circuit was in operation?

Chapter 22

Communications Regulations and Services

ITU Headquarters in Geneva, Switzerland.

Contents

In this day and age, we are used to developments in technology resulting in dramatic advances in products and services. We observe the capabilities of personal computers, entertainment equipment, automobiles and medical devices increasing exponentially each year. We expect to see new products announced as fast as we get comfortable with the existing ones and it's a race to see if we can remove the packaging before an item becomes obsolete.

Most such products are driven by a combination of technology and market forces. Telecommunications shares those drivers but also has an additional major driver—regulations.

Telecommunication Regulation

Telecommunications providers often attempt to make use of the latest advances in technology. But they often find themselves limited in what they can offer as products due to constraints imposed by government regulation. They can only offer the services that are allowed under the terms of various regulations. This regulation comes about in two forms:

- *Regulation of the radio spectrum.* This one effects radio the most. Early in the last century, governments realized that radio, by its nature, went across borders. Thus, to avoid chaos, international agreements were required to avoid interference between users in different countries even while complying with their own national regulations.
- *Regulated Monopoly Providers.* Most US companies are required to compete in an open market for customers and business and their success is determined by their efficiency and by how well they deliver their products and services. Violations of this principal can be dealt with harshly by various federal anti-trust laws.

In the US, there is a special category of company that is *allowed* to be a monopoly—a *regulated common carrier*. In the case of telephone companies the government realized that it didn't make sense for a number of companies to each install poles down every street to provide competitive telephone service. Instead, a single company would provide such service. But to avoid being able to take advantage of their monopoly, this company would agree to be a regulated carrier. A commission at the state or federal level, depending on whether the service is interstate (between states) or intrastate (within a state), analyzes costs and determines what rates could be charged. This arrangement has worked well for many years.

While the two main categories of regulation occasionally overlap, in this book we will focus on the radio spectrum issues, recognizing that from time-to-time the common-carrier issues will arise.

A Short History of International Regulation

By 1855, wire-telegraph providers were offering commercial service throughout much of the world. Lines did not cross borders because each country operated using its own standards. Representatives from twenty European countries signed the International Telegraph Convention in Paris in 1865 to establish common standards facilitating cross-border communication. They also established the International Telegraph Union (ITU) to serve as an ongoing body to maintain common operation and to deal with the standardization of newly developed technology.

Emerging technologies, such as the telephone and radio, expanded the breadth of the ITU and it changed its name to the International Telecommunications Union. Following the formation of the United Nations in 1948, the ITU became a special agency of the UN and moved from Paris to Geneva, where it remains today.

A notable success of the ITU is the fact that today, if you are willing to dial enough digits, you can automatically have access to virtually any network connected telephone on the planet from any other telephone!

What Does the ITU Do?

The ITU functions are distributed among three sectors:

- The ITU Telecommunication Standardization Sector's mission is to ensure an efficient and on-time production of high-quality standards (Recommendations) covering all fields of telecommunications. These Recommendations generally become standards applied by national organizations such as the American National Standards Institute (ANSI) in the US. Using common standards, equipment can work in different environments and signals can meaningfully cross national boundaries.
- The ITU Telecommunication Development Sector has well-established programs of activities to facilitate connectivity and access, foster policy, regulatory and network readiness, expand human capacity through training programs, formulate financing strategies and enable enterprises in developing countries. This sector tries to find ways to facilitate economical offerings in areas without services, for example.
- The ITU Radiocommunications Sector's mission[1] is to "maintain and extend international cooperation among all the Member States of the Union for the improvement and rational use of telecommunications of all kinds." In terms of radio regulation and service offerings, this sector has the most impact and is the one we will find most interesting. They define their

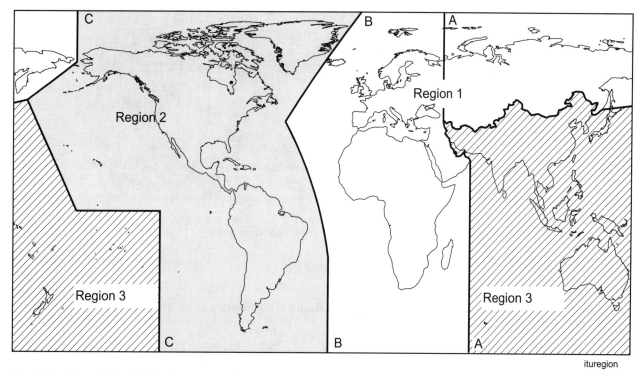

Fig 22-1—Map of the ITU Radiocommunications regions. [*Courtesy International Telecommunications Union.*]

primary focus as:

- To define the allocation of bands of the radio-frequency spectrum, the allotment of radio frequencies and the registration of radio-frequency assignments and of any associated orbital position in the geostationary satellite orbit in order to avoid harmful interference between radio stations of different countries;
- To coordinate efforts to eliminate harmful interference between radio stations of different countries and to improve the use made of radio frequencies and of the geostationary-satellite orbit for radiocommunication services. By so acting, they intend to ensure rational, equitable, efficient and economical use of the radio-frequency spectrum by all radiocommunication services, including those using satellite orbits, and to carry out studies and adopt recommendations on radiocommunication matters.

How Do They Do This?

The ITU has divided up the world into three *radiocommunications regions*, generally by North-South slices. ITU Region 1 includes Europe and Africa; ITU Region 2 is North and South America; and ITU Region 3 includes Asia and Australia. This is shown in **Fig 22-1**. Frequency allocations are generally aligned by ITU region, although many are common across multiple regions if communication can propagate between regions. The agreements result in treaties agreed to and signed by representatives of all the member countries.

All radio frequencies are specified as to the type of services they can be used for, essentially "from dc to daylight." In reality, the frequencies range from at least 9 kHz to 275 GHz, as of the latest references. For services that are fairly general, and fairly local, the agreements are clearly specified, but are left to national governments to administer within the regional boundaries. As an example, the frequency range from 315 to 325 kHz is assigned to "aeronautical radiolocation" in Region 1, "maritime radiolocation" in Region 2 and "aeronautical radiolocation and maritime radiolocation" in Region 3. That is a straightforward example.

In some cases, the frequency ranges are not the same in each region. I have compiled a shortened list of allocations for ITU Radiocommunication Region 2, based on various sources and have included it as **Table 22-1**. I have covered the VLF, LF and MF ranges and portions of the ELF and HF ranges in this table. I also grouped together some of the service types in Table 22-1 to make it reasonable in size, but this should provide an idea of the scope of international agreements.

For shortwave broadcasters the ITU maintains the function of coordinating frequencies and schedules between different users— often governments with competing interests in the use of the same frequencies. The ITU also administers orbital slots for geosynchronous satellites.

Regulation and Services at the National Level

Each ITU member nation has the responsibility to administer its treaty obligations in its area of responsibility. In the US, for example, the Federal Communications Commission (FCC), an agency of the

Executive Branch of government, has that responsibility. The FCC has a number of responsibilities (see **Fig 22-2**) related to the regulation of radio-based and other services:

- Determination of requirements for operator licenses for different services. For example, operator licenses are required for those who operate some types of shipboard radio systems, aircraft radio systems, Amateur Radio transmitters, mobile transmitters and broadcast transmitters. The FCC either prepares and conducts the

Fig 22-2—The many FCC functions are captured in this composite photo from their Web page.[2]

Table 22-1

Major Telecommunications Services Delivered by Radio (ITU Region 2) up to 10 MHz.

Frequency Range	Service	Notes
9-19.95 kHz	Radionavigation/Fixed/Maritime	Often long-range military strategic communications.
19.95-20.05 kHz	Standard Frequency and time signals	National Institute of Standards and Technology (US).
20.05-415 kHz	Mix of fixed, maritime and radionav	
415-510 kHz	Maritime mobile	Generally ship-shore radiotelegraph. Includes 500 kHz distress freq.
510-535 kHz	Aeronautical radionavigation	
535-1705 kHz	AM broadcasting	
1.705-1.8 MHz	Fixed/mobile/navigation	Note transition from kHz to MHz.
1.8-2.0 MHz	160 meter amateur	
2.0-2.3 MHz	Mobile, mainly maritime	
2.3-2.495 MHz	Broadcasting	120-meter international shortwave broadcast band.
2.495-2.505 MHz	Standard Frequency and time signals	National Institute of Standards and Technology (US)
2.505-3.2 MHz	Fixed mobile and aero mobile	
3.2-3.4 MHz	Broadcasting	90-meter international shortwave broadcast band.
3.4-3.5 MHz	Aeronautical mobile	
3.5-4.0 MHz	75/80 meter amateur	
4.0-4.65 MHz	Maritime mobile, fixed	
4.65-4.75 MHz	Aeronautical mobile	
4.75-4.995 MHz	Fixed, mobile (not aero)	
4.995-5.005 MHz	Standard Frequency and time signals	National Institute of Standards and Technology (US).
5.005-5.480 MHz	Fixed, mobile (not aero)	Includes five SSB channels for amateur use in US.
5.480-5.730 MHz	Aeronautical mobile	
5.730-5.9 MHz	Fixed, mobile (not aero)	
5.900-6.2 MHz	Broadcasting	49-meter international shortwave broadcast band.
6.200-6.525 MHz	Maritime mobile	
6.525-6.765 MHz	Aeronautical mobile	
6.765 7.0 MHz	Fixed	
7.0-7.3 MHz	40 meter amateur	
7.3-7.35 MHz	Broadcasting	41-meter international shortwave broadcast band.
7.35-8.100 MHz	Fixed, land mobile	
8.100-8.815 MHz	Fixed, maritime mobile	
8.815-9.04 MHz	Aeronautical mobile	
9.04-9.4 MHz	Fixed	
9.4-9.9 MHz	Broadcasting	31-meter international shortwave broadcast band.
9.9-9.995MHz	Fixed	
9.995-10.005 MHz	Standard Frequency and time signals	National Institute of Standards and Technology (US).

examinations themselves at field offices, or delegates the responsibility to others. Licensed volunteer radio amateur operators administer exams for Amateur Radio licenses under the auspices of approved Volunteer Examination Coordinator agencies (VEC), including the ARRL.

- Operator licenses of appropriate classification are required to adjust or repair most types of radio transmitting equipment. Many categories of licenses are available in different classes with progressively more challenging questions asked on the exam. Typically, the lower class licenses are for operators who only need to understand the rules so they can properly talk into a microphone without breaking any rules. Higher classes of licenses require testing more on technological aspects of the communication arts, sometimes with special test elements for equipment specific endorsements, such as radar. These are required for those who need to understand what goes on within the equipment so they can design systems around it or repair it.
- The FCC makes channel assignments within the ITU bands of frequencies for particular services. For example, in VHF commercial mobile services, public service mobile service, broadcast radio and television services, the assignments are on an individual user basis and include permits for transmitting station construction to specific standards of antenna height and EIRP. In other cases, the channels are assigned to a general user community for particular purposes. **Table 22-2**, for example, shows a list of HF marine channels for boat and shore station use. Note that while the US FCC assigns these individual channels, they are in line with the allocation of frequencies to this service in Table 22-1. In a similar manner, **Table 22-3** lists VHF marine channels for use in US coastal waters. Each marine radio generally comes equipped to

Table 22-2

A sample of HF Single Sideband Maritime Radiotelephone Channels.

ITU Channel No.	Coast Transmit (kHz)	Ship Transmit (kHz)	ITU Channel No.	Coast Transmit (kHz)	Ship Transmit (kHz)
4 MHz Duplex Channels.			**8 MHz Duplex Channels.**		
401	4357	4065	801	8719	8195
402	4360	4068	802	8722	8198
403	4363	4071	803	8725	8201
404	4366	4074	804	8728	8204
405	4369	4077	805	8731	8207
406	4372	4080	806	8734	8210
407	4375	4083	807	8737	8213
408	4378	4086	808	8740	8216
409	4381	4089	809	8743	8219
410	4384	4092	810	8746	8222
411	4387	4095	811	8749	8225
412	4390	4098	812	8752	8228
413	4393	4101	813	8755	8231
414	4396	4104	814	8758	8234
415	4399	4107	815	8761	8237
416	4402	4110	816	8764	8240 (USCG Calling)
417	4405	4113			
418	4408	4116	817	8767	8243
419	4411	4119	818	8770	8246
420	4414	4122	819	8773	8249
			820	8776	8252
421	4417	4125 (Calling, distress & safety, working on 4125 kHz simplex)	821	8779	8255 (Calling)
422	4420	4128	822	8782	8258
423	4423	4131	823	8785	8261
424	4426	4134 (USCG Calling)	824	8788	8264
			825	8791	8267
425	4429	4137			
			826	8794	8270
426	4432	4140	827	8797	8273
427	4435	4143	828	8800	8276
428	4351	(varies)	829	8803	8279
429	4354	(varies)	830	8806	8282
6 MHz Duplex Channels.			831	8809	8285
			832	8812	8288
601	6501	6200 (USCG Calling)	833	8291	8291
			834	8707	(varies)
602	6504	6203	835	8710	(varies)
603	6507	6206			
604	6510	6209	836	8713	(varies)
605	6513	6212	837	8716	(varies)
606	6516	6215 (Calling, distress & safety, working on 6125 kHz simplex)			
607	6519	6218			
608	6522	6221			

communicate on all of them and an individual user selects a channel for a particular purpose. Note that while these channels are assigned by the US FCC, and are generally in line with the assignments of most other countries, there are exceptions. The channels with an "A" suffix, for example are only used in the US. Outside the US these channels use separate ship and shore frequencies, instead of the simplex arrangement defined for US operation.

- The FCC establishes the technical and operational standards for all transmitting equipment and for many categories of users. These cover the range from technical issues (such as the level of spurious emissions allowed) to content issues—for example, specifying what is an obscene word or image that is not allowed to be broadcast!

- The FCC is responsible for the enforcement of its standards and rules regarding radio operations. To that end, it maintains a number of monitoring stations around the country, these can observe transmissions at most radio frequencies. In cooperation with other monitoring stations, they can triangulate to determine the location of an offending station. The FCC also has the responsibility to note and report on violations of treaty obligations if they interfere with US assignments.

Table 22-3

US Marine VHF radiotelephone channels assignments.

Channel Number	Ship Transmit MHz	Ship Receive MHz	Use
01A	156.050	156.050	Port Operations and Commercial, VTS.
			Available only in New Orleans / Lower Mississippi area.
05A	156.250	156.250	Port Operations or VTS
			Houston, New Orleans and Seattle areas.
06	156.300	156.300	Intership Safety
07A	156.350	156.350	Commercial
08	156.400	156.400	Commercial (Intership only)
09	156.450	156.450	Boater Calling. Commercial and Non-Commercial.
10	156.500	156.500	Commercial
11	156.550	156.550	Commercial. VTS in selected areas.
12	156.600	156.600	Port Operations. VTS in selected areas.
13	156.650	156.650	Intership Navigation Safety (Bridge-to-bridge).
			Ships > 20 m length maintain a listening watch on this channel in US waters.
14	156.700	156.700	Port Operations. VTS in selected areas.
15	—	156.750	Environmental (Receive only). Used by Class C EPIRBs.
16	156.800	156.800	International Distress, Safety and Calling.
			Ships required to carry radio, USCG, and most coast stations maintain a listening watch on this channel.
17	156.850	156.850	State Control
18A	156.900	156.900	Commercial
19A	156.950	156.950	Commercial
20	157.000	161.600	Port Operations (duplex)
20A	157.000	157.000	Port Operations
21A	157.050	157.050	U.S. Coast Guard only
22A	157.100	157.100	Coast Guard Liaison and Maritime Safety Information Broadcasts.
			Broadcasts announced on channel 16.
23A	157.150	157.150	U.S. Coast Guard only
24	157.200	161.800	Public Correspondence (Marine Operator)
25	157.250	161.850	Public Correspondence (Marine Operator)
26	157.300	161.900	Public Correspondence (Marine Operator)
27	157.350	161.950	Public Correspondence (Marine Operator)
28	157.400	162.000	Public Correspondence (Marine Operator)
63A	156.175	156.175	Port Operations and Commercial, VTS.
			Available only in New Orleans / Lower Mississippi area.
65A	156.275	156.275	Port Operations
66A	156.325	156.325	Port Operations
67	156.375	156.375	Commercial Intership
			Used for Bridge-to-bridge communications in lower Mississippi River only.
68	156.425	156.425	Non-Commercial
69	156.475	156.475	Non-Commercial
70	156.525	156.525	Digital Selective Calling
			(voice communications not allowed)
71	156.575	156.575	Non-Commercial
72	156.625	156.625	Non-Commercial (Intership only)
73	156.675	156.675	Port Operations
74	156.725	156.725	Port Operations
77	156.875	156.875	Port Operations (Intership only)
78A	156.925	156.925	Non-Commercial
79A	156.975	156.975	Commercial. Non-Commercial in Great Lakes only
80A	157.025	157.025	Commercial. Non-Commercial in Great Lakes only
81A	157.075	157.075	U.S. Government only - Environmental protection operations.
82A	157.125	157.125	U.S. Government only
83A	157.175	157.175	U.S. Coast Guard only
84	157.225	161.825	Public Correspondence (Marine Operator)
85	157.275	161.875	Public Correspondence (Marine Operator)
86	157.325	161.925	Public Correspondence (Marine Operator)
AIS 1	161.975	161.975	Automatic Identification System (AIS)
AIS 2	162.025	162.025	Automatic Identification System (AIS)
88A	157.425	157.425	Commercial, Intership only.

We all are familiar with some of the services that come to us by radio. Perhaps we are most directly affected by the broadcast services that entertain or inform us on a daily basis. These services are but a small fraction of all the communication services that benefit the world's people. Table 22-1 provides a summary list of the major services listed by frequency. I will briefly describe the key services so you know what the regulatory jargon refers to:

- *Aeronautical mobile* is a term referring to radio stations aboard aircraft. A number of frequency bands are provided. A VHF band is allocated for AM communications between aircraft and fixed base operations (FBO). This band is also used for announcing an aircraft's position while at or flying near an airport, and on occasion between aircraft. Included in this band is 121.5 MHz, the emergency frequency on which any pilot can transmit. This frequency is monitored by Air Traffic Control, Flight Services and by other pilots. In addition, the FAA's Federal Aviation Regulations require HF communication capability once an aircraft is 30 minutes or 100 miles offshore, for instance, on transoceanic flights. Other HF frequencies are allocated for airline companies to communicate scheduling information, maintenance issues or weather data. Ground facilities in support of aircraft can be licensed to use some frequencies within the aeronautical mobile allocations. Pilots with Amateur Radio licenses can communicate with other amateurs over long distances because of their altitude.[3]
- The *Amateur Radio Service* has frequencies in many bands throughout the MF, HF, VHF, UHF and microwave portions of the spectrum for use by licensed Amateur Radio operators. These

frequencies are provided to help advance the radio arts through non-commercial development and experimentation. Amateur Radio operators also provide communications in emergencies when commercial facilities are unavailable or overcrowded. They also simply enjoy making use of radio as a matter of personal interest.
- *Broadcasting* refers to informational or entertainment transmissions intended for multiple unspecified listeners. Included are voice broadcasts in the MF (typically national or regional) and HF (typically international in nature) ranges, as well as VHF and UHF television broadcasting.
- *Fixed station* refers to a radio communication station (which is not a broadcasting station) operating from a fixed location. This fixed station could be used to communicate between multiple fixed locations, as between a offshore oil platforms and a headquarters location, or between the fixed station and a service aircraft or ship. In the US, most land-based, fixed communications can be carried out by terrestrial telephone. However, in other parts of the world the infrastructure may not be available and HF radio between fixed stations is often the best solution for distances beyond line-of-sight.
- *Maritime mobile* is a term referring to radio stations aboard boats and ships. A number of frequency bands are available. A VHF band is allocated for FM communications between nearby craft, and between craft and USCG or other coastal or nearby emergency services. In addition, drawbridge, canal lock operators, port operators, marinas and yacht clubs may have shore station licenses to use specified marine VHF frequencies to communicate with boats and ships. As with aeronautical mobile, there are HF channel assignments

for long distance maritime communications between ships and shore facilities. In addition, maritime operators have radiotelegraph frequencies in the MF range. Large commercial vessels can use these, although radiotelegraph is no longer required for ships of many nations. Maritime users on the high seas have access to satellite communications using the INMARSAT satellite system. A special category of shore station is the *public correspondence* station. Public correspondence stations are licensed in both HF and VHF ranges and allow fee-paid connection between the radio link and ground-based telephone services.
- *Radiolocation* refers to determining the location of another station or object. Included are radar systems and some transponder systems. These will be covered in more detail in Chapter 23.
- *Radionavigation* refers to radio-based systems for position location. While designed primarily for aircraft and maritime users, the global positioning system (GPS), a satellite-based navigation system, has become popular for land-based navigation as well. In addition to GPS, there are a number of other terrestrial-based systems providing similar service. These will be covered in more detail in Chapter 24.
- *Standard Frequency and Time Signals* are transmitted by government agencies to propagate standard information. In the US, the National Institute of Standards and Technology transmits signals at precise frequencies of 60 kHz (WWVB) as well as 2.5, 5, 10, 15 and 20 MHz from station WWV in Fort Collins Colorado. The station radiates 10,000 W on 5, 10, and 15 MHz, and 2,500 W on 2.5 and 20 MHz. Although each frequency carries the same information, multiple frequencies are used so that propagation on at least one

frequency is available at all times. WWVH is a similar station located in Hawaii that broadcasts on the same frequencies as WWV, except for 20 MHz. In addition to standard radio frequencies, the signal format provides accurate audio frequencies, accurate time indicators and information about atmospheric conditions.[4]

For each category of service, there are generally associated requirements on modulation types, allowed bandwidth, frequency accuracy and stability, spurious emissions, etc. Each service has equipment requirements that manufacturers must meet and certify.

Notes

[1]ITU Radio Sector mission statement, **www.itu.int/ITU-R/ information/mission/index.html**.

[2]See **www.fcc.gov**.

[3]Thanks to private communication from Rosalie White, K1STO, an FAA licensed general aviation pilot and FCC licensed Amateur Radio operator.

[4]See **tf.nist.gov/timefreq/stations/ wwv.html**.

Review Questions

1. Why is it necessary to have an international body make and/or register assignments of frequencies for use by international shortwave broadcast stations? What would happen if a central agency didn't do that?

2. Repeat question 1 for satellite orbital assignments.

3. What must the FCC do if it wishes to make an assignment contrary to ITU allocations?

Chapter 23

Radiolocation Systems

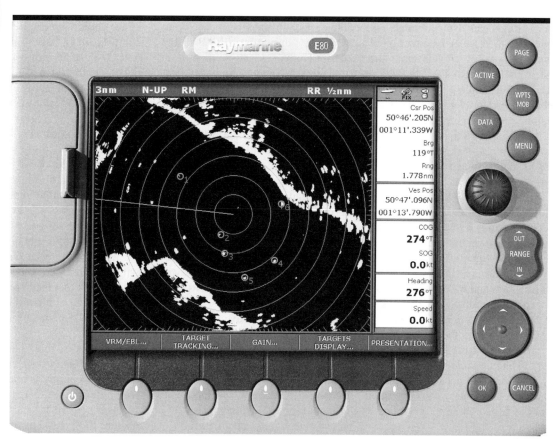

Display screen of Raymarine E-80 Multifunction Network Display showing land contours and selected targets. (*Photo courtesy Raymarine Inc*)

Contents

By *radiolocation* we generally mean a system for the determination of the location of another object. This is in contrast to a *radionavigation* system, where users determine their own location. We will discuss radionavigation in the next chapter.

An important radiolocation system is *radar*, short for RAdio Detection And Ranging, both of which can be determined by radar. Modern ground radar systems can also determine a target's altitude, velocity, direction of travel and possibly provide some idea of size. Radar is also used aboard aircraft for multiple purposes, including terrain avoidance, weather mapping and avoidance, fire control and surveillance. Shipboard radar is similar in many ways, but tends to be used for collision avoidance and navigation, as well as for fire control for military vessels.

In addition to radar, radiolocation includes various transponder systems, in which a station transmits an interrogation signal to a target and the target returns a signal with identification and other codes. In the military, such transponder systems are called *Identification Friend or Foe* (IFF), while in the civilian aircraft environment they are referred to as *secondary radar systems*. The key distinction, in my mind, is that radar does not depend on cooperation from the target, while transponder systems do.

Where Did Radar Start?

What we know today as radar was envisaged by radio pioneers in the early days of radio. They apparently noticed that signals were reflected from large metallic objects, and they proposed ship collision-avoidance-systems based on these observations. In 1904, a German engineer named Hulsmeyer obtained a patent for a radio-based "obstacle detector." His patent[1] was entitled "Hertzian-Wave Projecting and Receiving Apparatus

Adapted to Indicate or Give Warning of the Presence of a Metallic Body Such as a Ship or Train in the Line of Projection of Such Waves." This sounds a lot like radar, but unfortunately, the early radio art had not progressed enough to support the requirements of a serious radar system.

The approach of World War II was the impetus for a number of national defense establishments to develop a workable radar design. The British "Chain Home" radar system, a rudimentary HF system in today's terms, was remarkably successful at determining when enemy bombers were crossing the English Channel. It was a key element in the British success in the "Battle of Britain" that set the stage for the allies winning World War II.

Other radar systems evolved during that war, including a long-range search radar successful at detecting the attack on Pearl Harbor that brought the US into World War II. Unfortunately, the detections were misidentified as friendly forces by the Sunday morning staff—otherwise many events might have turned out differently. Another key development was the coupling of radar target definition with naval gun systems, allowing automatic aiming of weapons. In addition, special vacuum tubes able to generate microwave signals enhanced the capability of radar systems for all applications.

How Does Radar Work?

The conceptual design of radar is pretty straightforward. You send a signal towards a target and some of the signal gets reflected back. If you know where your antenna is pointing, you know the direction of the target. If you measure how long it takes for the signal to get there and back, you can compute the distance to the target.

Nothing hard about this, is there? Well, as with all great ideas, the devil is in the details. I will go into some of these details to give you an idea about how radar works. But beware, there's a lot more. Many very talented people have spent a great deal of time and effort developing some very sophisticated radar concepts that won't fit into this small chapter! I will try to describe some of the choices an expert radar engineer might have to make. This will give you an idea why not all radars are the same.

The statement above "*If you know where your antenna was pointing, you know the direction of the target*" should raise a question about antenna beamwidth. If your antenna has a beamwidth of 50°, you may have a problem deciding in which direction your target really is located.

If you have several targets in the same direction, one behind the other, the width of the radar pulse you send out will determine whether or not you can tell that they are both there. A long pulse width will blur signals together as though they were only one target.

You must be able to detect whatever signal is reflected from the target. This ability is related to transmitter power, antenna gain and receiver performance. I will discuss each of these concepts in more detail in the sections that follow. Again, we will be discussing basic concepts here. Recent advances in computation and signal processing have pushed modern radar performance and features well beyond the simple models I will discuss here.[2]

Heading Determination

Determining the direction of a target is largely a function of the beamwidth of the antenna. A narrow antenna beam will illuminate a small angular sector and targets that are returned must therefore be located within that sector. Of course, no antenna will have a beamwidth that immediately terminates at a particular angle; instead, the return from a target will be reduced as the beam moves off the target.

The beamwidth of an antenna is a function of the antenna design. For a parabolic antenna, the beamwidth becomes narrower as the parabola's diameter increases, for a given wavelength, or conversely as the wavelength gets smaller. Thus early lower-frequency radars were limited in angular resolution, for reasonably sized antennas. The advent of microwave transmission was a boon to radar from an angular-determination perspective. The half-power beamwidth, θ, of an optimally designed parabolic antenna can be approximated as: $\theta = 70\lambda / D$, where λ is the wavelength and D is the parabola diameter, both in the same units.[3] Note that for the parabola to be effective, D must be greater than λ.

I have calculated the beamwidth for a number of frequencies and parabolic antenna diameters and presented them in **Table 23-1**. The antenna diameters range from one small enough to be in the nose of an aircraft to one of the very largest, perhaps appropriate for a long-range radar searching for warheads in space. That information can be used with **Table 23-2** to determine how wide a field that beamwidth translates to as a function of range. This will give a sense of how finely you can discern targets in azimuth. In some cases, such as missile warning, it may be enough just to know whether or not there's one (or more) coming your way. It is a much more difficult problem to determine where each of those 1000 warheads is heading.

As with most aspects of radar engineering there are techniques that allow a better determination of angular position than just that defined by the antenna's inherent beamwidth, but that was the fundamental constraint facing the early radar designers.

Range Resolution

A typical radar sends out a short pulse of RF energy, followed by a longer period during which it listens for reflected pulses. In a typical system, the time between pulses is based on the maximum range. The time it takes a pulse to get from the transmitting antenna to the target and back to the receive antenna at the speed of light is just $\theta = 2R/c$, where R is the range and c is the speed of light. For example, if we want to have a radar with a 300-statute-mile range (see **Fig 23-1**) we would set T to at least $2 \times 300/186{,}000 =$ 0.003225 seconds, or 3.23 msec. This is the minimum *pulse repetition interval* (PRI) for that range. If we set the PRI to a smaller value, we would have multiple returns from a target within our range and there would be ambiguity in the distance from the antenna. This would appear in the form of multiple targets.

Once we have the PRI set, we can focus on the pulse width. For a rectangular pulse, the signal will start

Table 23-1

Parabolic antenna beamwidth (degrees) as function of frequency and diameter.

		Diameter (m)				
Freq (MHz)	λ (m)	0.5	1	10	30	100
100	3	21	7	2.1		
500	0.6		42	4.2	1.4	0.42
1000	0.3	42	21	2.1	0.7	0.21
5000	0.06	8.4	4.2	0.42	0.14	0.042
10000	0.03	4.2	2.1	0.21	0.07	0.021

Table 23-2

Target resolution (feet, *miles*) as a function of beamwidth (θ) and range.

θ (deg)	10	50	100	200	500	5000	Miles
0.01	9	46	92	184	461	4,605	
0.1	92	461	921	1,842	4,605	*9*	
0.5	461	2,303	4,605	9,211	*4*	*44*	
1	921	4,606	9,212	*3*	*9*	*87*	
5	4,617	*4*	*9*	*17*	*44*	*437*	
10	9,305	*9*	*18*	*35*	*88*	*881*	

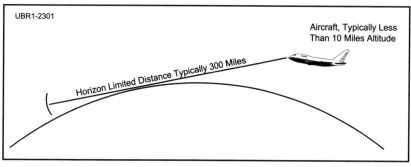

Fig 23-1—Operating environment for typical ground-based search radar.

to return from the target as soon as the leading edge hits the target, and continue until the pulse is past the target. Note that the receiver's internal filters can stretch the pulse up to twice its original length. If a second target's return overlaps the end of the first target's return, as shown in **Fig 23-2**, the result will be a single, wide pulse. The fact that there are actually two targets will not be apparent from the display. This is shown in a simulated oscilloscope display in **Fig 23-3**. In the absence of special waveforms and associated signal processing (called *pulse compression*) the range resolution is just equal to the pulse width divided by the speed of light.

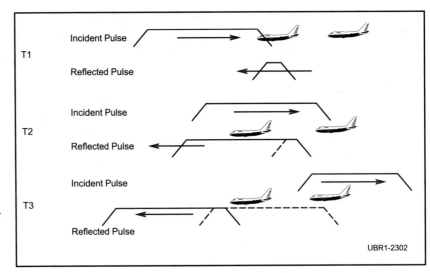

Fig 23-2—Illustration of range resolution issue with closely spaced targets. Dashed line is reflection from second target.

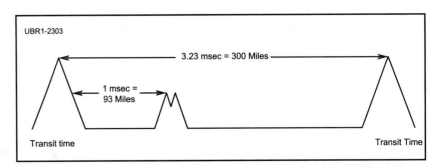

Figure 23-3—A search radar signal with two closely spaced targets 1 msec out.

How Do We Get the Signal Out There and Then Back?

The short answer is "send plenty and listen carefully"! The maximum range calculation is a function of all the radio parameters we've talked in earlier chapters: transmit power, transmit antenna gain, receive antenna aperture, receive sensitivity and the local noise level at the receiver. In addition, we have a new concept to deal with for radar—*equivalent reflecting area*, also known as *radar cross section*. We will use a ground-based, long-range, aircraft search radar as an example in the following discussion. This is a type of radar suited for Federal Aviation Administration air-route surveillance or air-defense use. Note that the same basic principles apply to other applications. Some radars will be designed for shorter ranges, or they will be designed to look for smaller targets with narrower resolution. Each will result in different parameter selections.

Step 1—Get the Signal to the Target

We transmit a pulse of RF towards the target once in each PRI. Typically the pulse is a very strong signal and is transmitted through a very high gain antenna. The *effective radiated power* (ERP) of such a system is often many orders of magnitude higher than for a communications system, but the same rules still apply. The transmitted signal can be expressed as the product of the peak power output during the pulse times the gain of the antenna (we will assume any transmission line losses are incorporated into the antenna gain to simplify the rest of this). In decibels, that is $P_{dBm} + G_{dB} = ERP_{dBm}$. The power density of the signal reaching the target is reduced by the distance, as we have observed and is just $ERP_{dBm}/4\pi R^2$, where R is the range to the target. The result is expressed in whatever units of distance we choose to use. This is just the same as we

have calculated for communications systems, in which the problem is usually solved at this point. The radar energy illuminates the target, but then most of it continues on. Compare the size of the beamwidths, for example, from Table 23-2 to the size of typical airborne objects and it will be clear that most of this energy is destined for travel through outer space!

Step 2—Send Some Back

Fortunately, some of the radar energy doesn't go past the target. Instead, it is reflected back. While some is reflected back towards the radar, most of the radar's energy goes every which way. The subject of *radiation scattering* from objects has been of interest to radar engineers since the beginning, and hopefully it will be enough for us to recognize that it is a complex problem. Whatever the object is that we are illuminating, there will be some energy headed back towards the transmitter, so that there is a signal strong enough to detect.

For the purposes of calculation, we can represent the surface as an equivalent area, as if that area had helpfully sent all the energy back towards us in the proper phase. This mythical *radar cross section* represents the sector of space that would reflect back the same signal as the combination of all the small pieces of the usually much larger aircraft we're looking for. So that equivalent area accepts the radiation heading its way and sends it all back towards the radar. Some folks commit their careers to minimizing the cross section, as in stealth aircraft designers, while others try to find ways to overcome such minimizations.

Step 3—Listen Carefully

The effective area sends us back a signal, what happens next? That signal propagates outward from the area in many directions, including

back towards the radar. The power density of the signal received then is just the reflected signal $ERP/4\pi R^2$, where R is now the distance from the target back to the receive antenna. For the remainder of this discussion, we will assume the usual *monostatic* case, that is the receiver and transmitter are in the same place and share the same antenna. The alternative *bistatic* configuration, with separate transmitter and receiver locations and antennas, is occasionally encountered, generally for special purposes.

In the monostatic configuration we note that the distance from transmitter to target is the same as from the target to the receiver, and thus the "R" distance in both the transmit and receive power density equations is the same. Knowing a bit about the receiver performance would now allow us to determine the answer to the important questions: "Can we detect the target?" or more helpfully: "How small a target can we detect?" Then there's: "How much power do I need to transmit to detect the size target I'm interested in at the maximum range?" Note that these are almost the same question, but with emphasis on solving for different parameters.

Put it All Together and What do you Get?

For the monostatic configuration we have noted that the range from radar transmitter to target is the same as from the target to the receiver. In addition, the receive antenna aperture is related to the transmitter antenna gain, since the same antenna is generally used for both transmitting and receiving. There are perhaps other obvious common elements, since the transmitter and receiver came from the same system designer. It's fair to assume that they both operate on the same frequency and that the receiver bandwidth is optimized to receive the pulse sent from the transmitter. We have defined the power density illuminating the target as:

$$P_T G_A / 4 \pi R^2 \qquad \text{(Eq 1)}$$

where P_T is the transmitted power and G_A is the antenna gain. The power density of the reflected signal at the receiver is:

$$P_R = ERP / 4 \pi R^2 \qquad \text{(Eq 2)}$$

where ERP is the effective radiated power of the reflected signal. But ERP is just the effective cross section, A, times the power density received at the target, or:

$$ERP = A P_T G_A / 4 \pi R^2 \qquad \text{(Eq 3)}$$

Substituting the above expression for ERP into the formula for P_R, we get:

$$P_R = A P_T G_A / (4 \pi R^2)^2 = A P_T G_A / (4 \pi)^2 R^4 \qquad \text{(Eq 4)}$$

This is significant. Note that unlike radio, where the signal falls off with the *square* of the distance; in radar the signal drops off at the *fourth power* of distance. This is why so much power and antenna gain are needed to detect distant radar targets.

If we incorporate the receive antenna aperture, the receiver sensitivity and the required SNR into Eq 4 above for received power density, we will have the famous *radar range equation*, a major aid in determining long-range radar performance. Let's take a quick pass at doing just that. You may need to develop this on a blackboard during a job interview someday. I did once!

Receive Antenna Aperture

In Chapter 17, we defined the relationship between antenna gain and effective aperture as:

$$A_{eff} = G_i \lambda^2 / 4 \pi \qquad \text{(Eq 5)}$$

where G_i is the power gain compared to an isotropic antenna, the same as the transmit power gain. Now that we know the aperture, if we multiply it by the received power density, we can determine how much signal we actually receive at the receiver as:

$$P_R = A P_T G_i / (4 \pi)^2 R^4 \qquad \text{(Eq 6)}$$

Received signal power at antenna terminals is:

$$(G_i \lambda^2 / 4 \pi) A P_T G_i / (4 \pi)^2 R^4 \qquad \text{(Eq 7)}$$

Note that, not surprisingly, the antenna gain shows up twice. Rearranging, and converting from wavelength to frequency, we have:

$$A P_T G_i^2 (c/f)^2 / (4 \pi)^3 R^4 \qquad \text{(Eq 8)}$$

Thus we neatly have the received signal power in terms of the effective target size, transmitted power, antenna gain, frequency and distance. Note that while we can easily calculate this result in terms of a fixed target size, in real life with a moving target we get a somewhat different effective size with each return, depending on the way the phase from each individual piece of the target adds up. This it's common to talk in terms of a range of values, usually expressing them as a probability function with an average and other statistical parameters.

So How Much Signal is Enough?

The last step in this development is to determine the signal to noise ratio (SNR) at the output of the radar receiver so we can tell whether or not we can detect a target of a certain effective cross section A. In previous chapters, we discussed the sources of receiver noise and how they effect sensitivity. With a radar system, we can assume that the internal receiver noise will dominate. This will not always be the case; for example, the antenna may swing past the direction of the sun or other strong noise source. But at other times the radar will be looking at a quiet sky, so we'll have to use that as our design point. If a large noise source is in the direction of the target, we will either risk not detecting a target or we will have to increase power.

I mentioned in the discussion of receivers that the noise output of a receiver is proportional to the bandwidth of the receiver, and that is a key parameter here. We haven't talked about pulse width since our discussion of range resolution, but this is where it shows up again. The narrower the pulse width, the wider the bandwidth required to properly receive it. At a first approximation, we can say that the optimum bandwidth is just equal to $1/\tau$, where τ is the pulse width. Thus, assuming optimum system design, we can express the bandwidth in terms of the primary radar parameter—that is, pulse width.

The receiver's internal noise is a function of the bandwidth and a parameter that we haven't discussed previously, the *effective system temperature*. By multiplying the effective absolute (degrees Kelvin) system temperature times Boltzman's constant (1.38×10^{-23}) times bandwidth, we get the noise power from the receiver. By the way, this can work for both internal and external noise sources, with external sources being evaluated as to effective temperature. This topic could generate a book of its own, so for the moment we'll leave it in general terms. The average (again we're talking statistics here) output noise power of the receiver in watts would thus be:

$$P_N = K T B \qquad \text{(Eq 9)}$$

where K is Boltzman's constant, T is the system absolute temperature and B is the bandwidth in Hz (typically, $1/\tau$, with τ the pulse width in seconds). Substituting $1/\tau$ for the bandwidth, we have:

$$P_N = (K T) / \tau \qquad \text{(Eq 10)}$$

Notice that we have calculated both the received signal power and the receiver noise power. If we divide the signal power by the noise power, we get the SNR:

$$SNR = [A P_T G_i^2 (c/f)^2 / (4 \pi)^3 R^4] / [(K T) / \tau] \qquad \text{(Eq 11)}$$

This is just what we were looking for. It's rather long, but we should be able to make sense of all the terms. **Table 23-3** presents two examples of the calculated *single-pulse* SNR in dB as a function of the parameters above. In most applications, we will have more than a single pulse to use, but this is where to start. I have chosen a frequency typical of long-range search radars (1 GHz) with a 10 μs pulse width and another, perhaps appropriate for shorter-range, higher-resolution shipboard use at a higher frequency (10 GHz) with a 2 μs pulse width. Note that while both sets of calculations assume an antenna gain of 1000 (30 dB), the 1-GHz antenna will be about 10 times the size of the 10-GHz antenna.

Table 23-3

Received signal-to-noise ratio of a single pulse as a function of peak transmitter power and distance.

Case 1, Frequency 1 GHz, receive temperature 293° C, Pulse width 10 μs, antenna gain 1000, cross section 10 m².

PeakDistance (miles)

Power (W)	1	10	50	100	200
100	72.2	32.2	4.3	−7.8	−19.8
1000	82.2	42.2	14.3	2.2	−9.8
10000	92.2	52.2	24.3	12.2	0.2
100000	102.2	62.2	34.3	22.2	10.2
1.00E+06	112.2	72.2	44.3	32.2	20.2

Case 2, Frequency 10 GHz, receive temperature 293° C, Pulse width 2 μs, antenna gain 1000, cross section 10 m².

PeakDistance (miles)

Power (W)	1	10	50	100	200
100	45.2	5.2	−22.7	−34.8	−46.8
1000	55.2	15.2	−12.7	−24.8	−36.8
10000	65.2	25.2	−2.7	−14.8	−26.8
100000	75.2	35.2	7.3	−4.8	−16.8
1.00E+06	85.2	45.2	17.3	5.2	−6.8

So what do we do with this returned signal? Early radar displays used a typical oscilloscope display with a fixed antenna to indicate the range to a target within the beam. The horizontal sweep would be set to synchronize with the PRI so that each sweep would show the full range of the radar and a target's progress up or down the beam could be observed, as shown in **Fig 23-3**. Radar folk would call that an "A-scope."

Let's look at some numbers. Assume we have a 3-msec PRI, meaning that about 330 pulses will strike the target every second. An aircraft moving at say 300 miles per hour will have moved 400 feet during that second, still within the typical range resolution (0.93 miles in this case). Thus the intensity of the display will be enhanced as the signals from multiple pulses tend to add together on the display, which uses a long-persistence phosphor in its CRT. This increases the apparent SNR by combining the display from one PRI with the next, as long as they are within the same effective *blip* on the screen. In other words, the scope's persistence displays the sum of all the returns as one slightly larger, but quite a bit brighter target.

Since signals from each return are not actually in phase, rather than adding linearly the successive returns add as about the square root of the sum of the individual pulses. If we add that up over that second, we get an integration gain of $(400)^{1/2}$ or 20, an additional 13 dB of SNR. This value can be added to the SNR of a single pulse shown in Table 23-3. On the other hand, a subtraction must be made for any additional losses in the system to determine the net SNR.

The SNR required for detection depends on the ability of the operator to find and declare a return on an analog PPI screen. An experienced and alert operator can observe a target with a very low SNR, while a digital system generally requires a fairly high detection threshold, perhaps 10 dB, to minimize the likelihood of false alarms.

Adding Another Dimension

It didn't take folks long to figure out that a radar that only showed targets in one particular direction was not all that helpful, at least for many requirements. The great leap forward was the *plan position indicator* (PPI). The PPI is arranged like a plotting board on a large circular screen. The outside edge of the circle represents the maximum radar range, the center of the circle the radar location, and the display rotates around the circle in synchronism with the rotating antenna. As the antenna beam moves past a target, its illumination would diminish, except that the display tube's persistence keeps the display intact until the next revolution of the antenna. A typical search radar system might rotate at a scan rate of 1 revolution per minute, or 360° in a minute. A sketch of a PPI is shown in **Fig 23-4**. PPIs were known as "B-scopes" to early radar operators.

A Few Challenges Remained

The development of radar during WWII was a major effort that took place on both sides of the Atlantic and by both sides in the conflict. That the collaborative work of large UK and US teams was much more successful may be one of the reasons for the ultimate outcome of that conflict. Many technological challenges were overcome in the process, and they were resolved much more quickly than would have been the case had not this been a high priority in both nations. Of particular note was the development of the *magnetron*, a special type of vacuum tube able to generate microwave energy and that could be easily produced in large quantities.

What About Bumblebees?

One interesting problem had to do with the nature of the fourth power

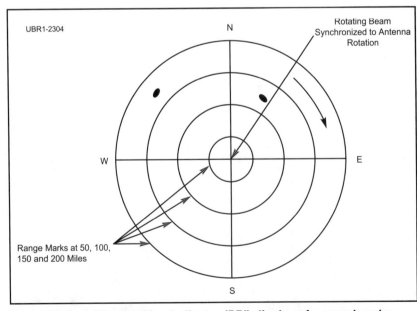

Figure 23-4—A Plan Position Indicator (PPI) display of a search radar output. Two targets are shown, one to the NE at about 120 miles and one to the NW at 170 miles.

relationship between range and received signal strength. As previously noted, this resulted in the need for very high transmitted power levels to obtain the desired range. The other side of that coin is that signals returned from nearby targets are very strong, so that close-in small objects have the effect of saturating the receiver and blinding it from seeing larger targets. It may be interesting to see a nearby swarm of bumblebees on the radar screen, but if it keeps you from seeing an aircraft at the same heading, it's not too helpful! The solution is to reduce the sensitivity of the receiver during the early portion of each PRI. In fact, if it is reduced by the same fourth power relationship, the intensity of a particular target display can be constant throughout the radar range. This improvement is known as *sensitivity time control* or STC.

Moving Versus Fixed Targets

A major limitation of early aircraft-detection radar was ground *clutter*. Reflected signals from structures and terrain features tended to mask returns from desired targets. The solution was to use a *moving target indicator* (MTI) processing system. The first MTI processors were composed of delay lines, equal in delay to the PRI. The returns from one sweep would be subtracted from the same signal delayed by exactly one PRI. In theory, anything that hadn't moved would be subtracted out and the remainder would be targets that had moved during the PRI.

A more successful system was deployed later using high-resolution analog-to-digital conversion techniques made possible with integrated circuits. The digitized signal would then be processed using multiple shift registers to provide the delay. The process could be repeated multiple times. In order to be successful, all aspects of the radar signal had to remain constant from pulse to pulse, or any change would look like a moving target.

A variation of this problem was apparent to users of shipboard radars, except it was caused by reflections from waves. Such *sea clutter* was reduced significantly by changing to horizontally polarized antennas that had a null at zero degrees of elevation.

Radar Applications

In addition to the search radars used as examples above, radar systems have a number of other applications.

Height-Finder Radar

The PPI display presents a two-dimensional view of the universe. Aircraft targets are presented in terms of the place on the ground that they are above, but no information is provided (by the radar) about their altitude. As we will discuss later, cooperating aircraft provide this information in a different way, however, if we are tracking hostile aircraft they cannot be expected to be so accommodating as to tell us their altitude!

A height finder is just a search radar laid on its side so the antenna goes up and down instead of back and forth or around. The operator can align the height finder to the azimuth of the desired target and determine the elevation angle by noting the maximum response from the antenna pattern. By knowing the angle and the hypotenuse of the triangle (from the search radar) we can compute the horizontal distance (very analogous to the search radar's range for distant targets) and the height above ground.

Fire Control

One of the major WWII applications of radar was as part of gun-pointing fire-control systems for naval guns. The radar provides target location data to the system that then incorporate inputs for the ship's speed, pitch and roll. These systems then compute the gun firing solution, including timing the firing to coincide with the appropriate roll position. Modern radars are also used in air, sea and ground systems to acquire targets and provide coordinates to systems that direct defensive missiles on an intercepting course.

Airborne Radar

Most larger aircraft include forward-facing radar located behind a radome in the nose of the aircraft. For civilian aircraft, the system's primary function is weather avoidance. In military aircraft, they also provide data to weapon-control systems.

Terrain Mapping Radar

Aircraft radar that is pointed forward and downward can be used to map terrain features. By making parallel multiple runs at the same altitude and recording and digitizing the data, an effective *radar map* can be generated. This can be used as data for a radar-based navigation system that can match observed radar data with the stored map to determine aircraft location and track.

Fig 23-5—Four foot open array radar antenna rated at 4 kW, a part of the Raymarine E-80 Multifunction Radar System (*Photo courtesy, Raymarine Inc*)

Ship Navigation Radar

Shipboard radar can be used to allow operators to see through fog as well as operate safely at night. The ship's position can be determined by the reflections from navigational buoys, most of which include a special corner-reflector on the top to maximize radar returns. In addition to buoys, ship radar can detect other vessels and surface structures. A modern shipboard antenna is shown in **Fig 23-5**. Note that the PPI configuration display shown in the photo at the front of this chapter shows both ship returns as well as coastal features, both of critical interest to those on duty on the ship's bridge.

Weather Radar

While weather cells can interfere with search radar operation, the reflections are just what's called for to locate and track rain or thunderstorm activity. The US National Weather Service uses a network of interconnected computers and radar systems to track weather systems and provide input to forecasting computers designed to provide warning of serious weather conditions.

Speed-Control Radar

Speed control radar is a somewhat different application of reflected radio waves. A speed-control radar emits a continuous signal at a fixed frequency. The frequency reflected by a moving object will be changed by the Doppler effect, increasing if the target is approaching, decreasing if moving away. By heterodyning the received signal with the transmitted signal, the resulting difference frequency will be the Doppler. The Doppler difference frequency is just equal to $2v/\lambda$ where λ is the signal wavelength and v is the object's velocity (towards or away from the observer). Both velocities are expressed in the same units. If the velocity is towards the observer, the difference adds to the frequency; if away it subtracts. Thus the velocity of the object can be determined by a simple calculation and displayed. The same technique can be applied to other radar systems, such as aircraft search radar to determine target velocity.

Modern Radar Systems

Modern radars operate on the same principles as the early ones from WWII; however, in all aspects radar has taken advantage of the advances in electronic and computational technology.

Solid-State Integrated Displays

Perhaps the most striking change is in display technology. The rotating beam of a cathode-ray tube based PPI screen has been almost entirely replaced by modern display technology. The flat panel display of a current marine radar is shown in the photo at the front of this chapter. In addition to the radar return data, other information is also presented, providing a single integrated display with all key information on the single display screen.

In addition to displays, microminiature solid-state digital electronics have replaced many of the cumbersome electromechanical systems of early radars. Digital electronics have also made possible the sharing of data between centers and coordination of radar-based information to an extent far beyond that provided by early manual systems.

Phased-Array Radar Systems

A major change in radar technology was developed by a number of companies in the 1960s. In place of moving antennas, they built radar system designs using a large number of relatively low-powered transmitter/receiver (transceiver) combinations, each with an antenna directly attached to it. The transceiver/antenna modules were then combined into one large flat surface. The key ingredients that made this into a remarkable system were high-speed computational capabilities and electronically controlled phase-shift networks. By rapidly shifting the relative phase of the signals from each of the transmitters, the direction of the wavefront leaving and being received from the combined antenna system could be shifted almost instantaneously.

A normal scanning mode could be synthesized during periods of search, or a selected target could be specifically tracked, while simultaneously scanning the full range of the radar for additional targets. Both azimuth and elevation information could be obtained from the same system. The lack of moving parts provided for improved reliability, and the inherent redundancy of the design allowed for minimally degraded operation if some of the individual elements failed. The only major limitation was an inability to scan very far off axis. However, in some cases this limitation was overcome using multiple systems. This system has been used in major air defense and shipboard missile control systems.

Radar systems are critical for many applications, especially those in which the target cannot be counted on to cooperate. This includes not only hostile aircraft, but weather systems and landmasses as well!

The second kind of radiolocation system involves a beaconing function. This was a system developed during WWII because radar got too good at finding aircraft and many friendly aircraft were erroneously engaged and shot down, since their radar blips looked the same as the enemy's radar blips. The first systems were called *identification-friend-or-foe* or IFF. An aircraft or ship observing a radar return would send an interrogation signal, around 460 MHz, to the target. The friendly target's IFF system would respond with a "squawk" signal including a highly guarded code word. If it were the "code word of the day," the interrogator would conclude the target was "friendly."

Modern beacon systems form a major part of the worldwide air traffic control (ATC) system. In place of the IFF code word, modern beacons transmit a code indicating the category of aircraft and the altitude measured by that aircraft. The ATC radar antenna typically has a beacon transmit antenna mounted on it so that it points in the same direction as the radar. The beacon receiver data is presented on the same integrated display screen with the radar data and instead of a radar blip the beacon target is shown with altitude and type code. A look at an ATC display will show very few "skin" (radar) prints but mostly beacon tracks.

Note that the beacon does not suffer the range-to-the-fourth-power problem that requires such a high ERP from a radar transmitter. A beacon signal falls off with the square of the distance, as does any other radio signal. Modern beacons operate in the GHz range.

Emergency Locating Beacons

A completely different type of radiolocation system is that known as *emergency locating transmitter* (ELT). There are two similar systems: ELT designed to locate aircraft that have crashed, and Emergency Position Indicating Rescue Beacons (EPIRBS) for boaters. The aircraft ELT uses automatic switches designed to activate when the aircraft hits the ground harder than usual, much like the switch in automobile airbag system.

Marine EPIRBs are usually activated manually, although some will activate automatically when immersed in water. They obviously must survive in the water. Both types of emergency beacons transmit simultaneously on the international distress frequencies of 121.5 and 243.0 MHz. These frequencies are monitored by satellite systems that can relay received beacon signals to ground stations. Search aircraft can also use the beacon signals to locate targets, once they are within range.

A newer system operates on 406 MHz and includes a registration number allowing responders to identify the type and user of the beacon to ensure appropriate response. Some EPIRBs include a GPS receiver and can automatically send their position, while others may be connected to external navigation systems.

Notes

[1] British patent 13,170.
[2] Thanks to retired MITre Corporation Consulting Engineer, former Raytheon radar system engineer extraordinaire, my mentor and long time friend, George W. Randig, W1WO, for reviewing this chapter.
[3] *The ARRL Antenna Book*, 20th Edition, ARRL, 2003, p 18-18.

Review Questions

1. Describe how radar might have made a big difference to the outcome of WWII. For example, what would be the impact on the number of interceptor aircraft required by Britain to oppose German bombing if they did not have warning radar?

2. Calculate the average SNR of a search radar operating at 2 GHz with a receive temperature of 293° C, a pulse width of 6 ms, an antenna gain of 2000 and a radar cross section of 5 m^2.

3. What frequency would be output from a Doppler radar at 10 GHz if you were approaching it at 75 mph (watch your units!)?

Chapter 24

Radionavigation Systems

Leroy Chiao, KE5BRW, set a record in space by making 23 ARISS contacts with Earthbound school students while heading the Expedition 9 crew aboard the International Space Station.

Contents

Where Are We?

By *radionavigation* we generally mean a system for the determination of the location of the user. This is in contrast to a *radiolocation* system of the last chapter, where we try to find the location of someone else. The two overlap to a certain extent. One reason for having a radar on a boat or ship is to be able to find navigation buoys or land masses at night or while in the fog to determine your position. In this chapter, I will concentrate on systems specifically designed for the purpose of radionavigation.

There are many radionavigation systems that are now in use or that have been used over the years. This chapter is not intended to be a comprehensive examination of radionavigation systems, but rather a summary of some of the most important types, with applications of such systems. We will look at some of the key types in the historical order in which they came into use.

Medium-Frequency Direction-Finding Beacons

The most straightforward radionavigation system is probably the *MF radio beacon*. This is a radio transmitter whose primary function is providing a stable signal from a known location and with just enough identification information so that a user can determine which one of the many beacons it is. MF beacons have been deployed since the early days of radio for both ship and aircraft use. They typically operate in the range of 200 to 400 kHz and emit a constant carrier modulated by low-speed Morse-code identifiers.

Manually Operated Direction Finders

In both ships and aircraft, the navigator can use a receiver with a moveable directional antenna coupled to a direction scale. A typical antenna is a small bidirectional loop coupled with a *sense* antenna. Aircraft, because of their metallic skin, typically had a loop antenna on the top of the aircraft with a receiver remotely located at the navigator's station. For aircraft and for metal-hulled ships, the remote antenna typically had a calibrated control at the navigation station labeled RADIO COMPASS that the navigator could use to rotate the antenna and observe the signal strength.

The small loop has a directional pattern very much like the dipole discussed earlier, although it has a much smaller effective aperture. See **Fig 24-1** for the pattern of a small loop. In typical use, the navigator on a boat would turn the antenna to maximize the signal level and tune the radio for optimum reception. He would then verify the station's identity and take a bearing using the sharp null of the antenna pattern. By using the sharp null, rather than the wide peak, a more precise reading could be taken. Under good conditions, the accuracy was comparable to taking a bearing with a hand compass, or even the ship's steering compass under visual conditions, usually within a few degrees.

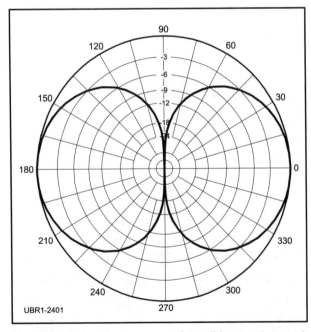

Fig 24-1—Azimuth response of small loop antenna at 300 kHz.

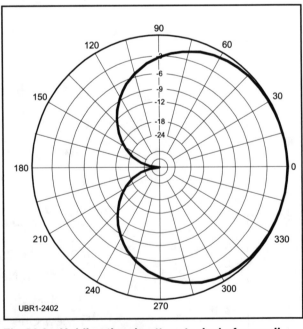

Fig 24-2—Unidirectional pattern typical of a small loop with a sense antenna.

A dipole or small loop has not one, but two nulls, 180° apart. Sometimes the correct direction is obvious. For example with a water edge beacon, one direction would be in the water, one on land. Sometimes the determination of which bearing is correct is not obvious, particularly in aircraft, and the *sense* antenna provides a way to determine the correct direction. The sense antenna is usually a small "whip" on a portable DF. An aircraft might use a wire from the fuselage to the top of the vertical stabilizer. The sense antenna's output can be adjusted in phase and amplitude and combined with the loop's output to form a unidirectional pattern, as shown in **Fig 24-2**. **Fig 24-3** and **Fig 24-4** show a military and a civilian version of manual direction finders.

Automatic Direction Finders

Instead of mechanically turning the loop antenna to determine the bearing, modern DF systems make use of a pair of perpendicular loops. By taking the outputs from each and combining them with variable phase, the pattern can be rotated without actually turning the antennas physically. Early systems used this technique with remote, transformer-coupled outputs in which the amount and phase of the coupling from each antenna could be manually adjusted to allow the navigator to remotely turn the antenna beam.

It wasn't long before modern electronics jumped in and used the two loop antenna outputs to provide signals to processors that could automatically compute the bearing to a beacon station. In the aircraft navigation environment, the beacons are called *non-directional beacons* (NDB), as contrasted to other systems described below that have signal coding that defines the direction.

Systems described below have largely superceded RDF systems, at least in the US. In both marine and aircraft environments, the US government is migrating from RDF. While some beacons have been decommissioned, the principles are still sound and the receivers can generally operate with AM broadcast signals as well as dedicated beacons, so the technology will remain useful for some time.

Fig 24-3—Military (WWII) manual radio direction finder. This version uses two loops to resolve the 180° ambiguity.

Fig 24-4—A commercial manual radio direction finder from the 1960s.

A hyperbola is a geometric figure with the property that any point on it is related to two points called *foci* (the plural of focus) so that the difference in distance between any point on the curve and two foci is the same. The idea is shown in **Fig 24-5**. The straight line midway between the foci is a special case in which the difference in distance is zero for any point on the line. There are an infinite number of hyperbolic curves between the foci, and each will have a different distance.

Imagine that we have a radio transmitter at each of the two foci and that they both transmit a pulse at exactly the same time. If we have a receiver that can measure the difference in the arrival time of the pulses, and we know where the transmitters are, we can calculate the hyperbola that we are on. Knowing that we are on a particular hyperbola doesn't tell us where we are, but does put us on that curve. If we avoid the region between the two transmitters, the hyperbola appears fairly straight and can act as what a navigator would call a *line of position* that can be drawn on a plotting chart. Now if we happen to have a

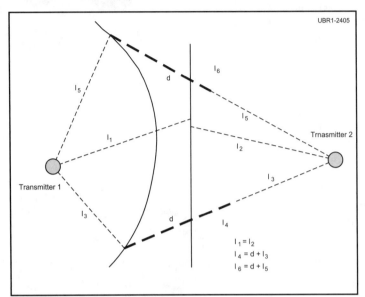

Fig 24-5—Illustrating geometry of a hyperbolic navigation system.

Fig 24-6—LORAN C receiver aboard sloop *Windfall* showing hyperbolic delay intervals from two stations compared to the master.

second pair of transmitters having hyperbolas that cross the first one, and we can determine which hyperbola we're on for the second set of hyperbolas, we have two intersecting lines of position. We are located at the point where they two lines intersect.

Enter the LORAN system

LORAN is an acronym for *LOng RAnge Navigation*. The concept was developed and was fairly successfully implemented during WWII. Instead of having a pair of transmitters synchronized together, a *chain* of typically five transmitters would be set up in a region, with one as the *master*. The master would send its coded pulse train and each slave station would send a pulse train as soon as the master's signal arrived at the slave station. The *time differences* (TDs) between each pair of stations would be different and they were spaced so that almost anywhere in the coverage region (typically greater than 1000 miles) there would be a pair of stations that would have almost perpendicular lines of position.

The early systems

used frequencies around 1.8 MHz, just above the AM broadcast band, and shared with the amateur 160-meter Amateur Radio band. This worked well in daylight, but night-time ionospheric propagation could cause changes in the delay that would make for inaccuracies. Early receive equipment was quite large and power hungry. The navigator would look at an oscilloscope and watch the pulses from the two selected stations on the same display. He would "crank in" delay with a calibrated adjustable delay network until the pulses lined up and would then read the delay difference from the delay network dial. Special navigation charts with lines of constant delay would be used to plot the ships position.

In the 1960s, the LORAN C system was deployed with transmitters operating at 100 kHz, a frequency that avoided ionospheric propagation difficulties. In addition, electronics were advancing to the point that circuitry could automatically perform the delay determination and the delay could be directly viewed on a digital display. By the early 1980s, LORAN receivers had become affordable for very small boats. In 1984, I purchased a LORAN receiver that weighed about two pounds and was the size of a small radio receiver. **Fig 24-6** is a photo of my receiver aboard the sloop *Windfall*, showing the delay differences, in msec from two transmitters. It not only could tell me the delays between transmitters, but also could convert directly to latitude and longitude if I needed to tell someone my position. Many nautical charts then included LORAN lines of position, and it was quite easy to determine position to within about 50 yards in my area.

Another feature was the ability to store *waypoints*. A waypoint is a set of LORAN coordinates for a place you had been. These could be entered by just hitting a button to memorize the location. The receiver could then compute the bearing and distance to a waypoint, as shown in the lower row of numbers in **Fig 24-7**. Returning to the waypoint was much more

Fig 24-7—LORAN C receiver of Figure 24-4 in navigation computer mode showing (bottom row) the course to steer and distance to waypoint number 1.

accurate than just finding your position on a chart, since any systematic errors (such as propagation delay over land versus water) would cancel out or be the same each time you were at that particular location. In other words, *repeatability* was better using the LORAN-C system than absolute *accuracy*. Using waypoints usually got me within about five to ten yards of any particular waypoint. I had to be careful to not memorize the exact location of a buoy as a waypoint, so that later I didn't hit it in the fog!

VHF Omnidirectional Range (VOR)

The *VHF Omnidirectional Range* or VOR system has been in place since around WWII, making obsolete a less-useful LF range system. VOR has been supported by the US Federal Aviation Agency (FAA) as a primary tool for aircraft navigation. There are a few different flavors—*Terminal VOR*, or TVOR, is a short-range system deployed around airports. The FAA supports longer-range systems for those below an altitude of 18,000 feet: *Victor Airways* or L(ow)VOR. Above 18,000 feet, *Jet Airways* or H(igh)VOR is part of the FAA air-route system.

VOR uses frequencies from 108 to 117.9 MHz, just below the aircraft communication range, making it possible for common antennas and radio equipment to support both functions. Channel spacing is 100 kHz. Aero navigation charts indicate the position and frequency of VOR transmitting stations.

The transmitting station broadcasts a signal with unidirectional pattern like that of Fig 24-2. The signal rotates through 360° at 30 revolutions per second. While this could be generated via two antennas out of phase by 90° with one rotating around the other, a more reliable system works in a manner similar to the ADF, without moving parts. A central monopole serves like the RDF sense antenna in combination with two pairs of phased bidirectional antennas positioned to result in a unidirectional pattern. The antenna elements are shown in **Fig 24-8C**.

In operation, the VOR drops its carrier momentarily when the peak of the beam is at magnetic north to allow synchronization between the transmitter and receivers. A measurement of the phase of the resultant 30-Hz amplitude modulation of the signal compared to the received phase at dropout provides a measurement of the direction from (or towards) the VOR transmitter. The aircraft can "dial in" a bearing and the aircraft instruments will indicate the direction the aircraft needs to turn to stay on the vector towards (or away) from the transmitter.

There are a number of other aircraft navigation systems using multiple beams with different modulating frequencies to indicate sectors that aircraft are in. These provide for *instrument landing system* (ILS) course determination and glide-slope indication.

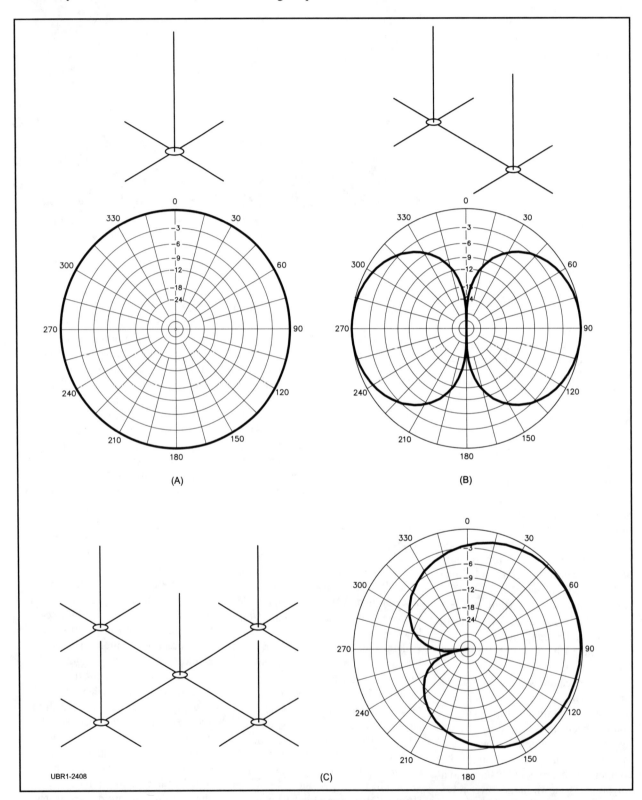

Fig 24-8 At A, omnidirectional VOR center antenna element with azimuth pattern. At B, a pair of bidirectional antenna elements with azimuth pattern. At C, combined VOR pattern at intermediate phase.

The Global Positioning System (GPS) is the most recent entrant to the "where are we" sweepstakes. GPS was put in place by the US Department of Defense as the *NAVSTAR* system in the late 1970s to provide position data to US military systems. In the 1980s, the US Government decided to use GPS as a replacement for the multitude of civilian navigation systems supported by the Department of Transportation. Most systems have phase-out plans in place, although there are so many users of RDF and LORAN systems, it has not proved to be an easy job to turn off these older systems. GPS is now fully available for civilian, as well as military use, and provides a level of accuracy beyond earlier systems. The combination of the world-wide availability of GPS with modern electronics technology has made GPS an essential element of many integrated telecommunications and navigation systems.

Like LORAN, GPS is a system based on time-difference measurements. GPS is a satellite-based system and depends upon a constellation of 24 low earth orbiting satellites operating at about half of the geosynchronous altitude. The satellites travel in six orbits with each satellite making two orbits per day around the Earth.

Each solar powered (with battery backup) satellite transmits a 50-W signal on a frequency of 1575.42 MHz containing a coded identification number, as well as information about its position (called *almanac data*) and information about the time of transmission and status (called *ephemeris data*) based on a very accurate on-board atomic clock. Corrections to the data in each satellite are provided by ground-control stations that send correction information to the satellites as needed.

If a ground station knows what time it is, by comparing current real time when receiving the message to the time that the satellite says it sent out the message, the receiver can determine the distance it is from the satellite. Because the satellite sends information on its exact position when the signal was sent, that means that the receiver is located on a sphere with the satellite at its center. A sphere 12,000 miles in radius doesn't make for a good position estimate, which is why signals from multiple GPS satellites are required to determine an accurate position. The signal from a second satellite defines a second sphere. The intersection of the two spheres places us on a circle. A third sphere, based on reading a signal from a third satellite will intersect the circle at two points. In many cases, one of the points is clearly not our location—it could be 24,000 miles from Earth, for example. In the worst case, a fourth satellite signal can get us to a single point in the universe, with latitude, longitude and altitude within about 50 meters. Auxiliary ground systems can eliminate additional errors and result in accuracy in the 5 to 10-meter range.

The trick is to know what time it is at the receiver, without having to buy a $50,000 atomic clock for every $99 GPS receiver! Obviously any inaccuracy in our time keeping will cause us to make errors in our distance calculations. Since the government put an atomic clock in each satellite and keeps the satellites in precise orbits, we can make use of satellite data to correct our own receiver's clock. All it takes is two or more satellites. If our clock is exactly right, the distance from the next satellite to our computed position will confirm that position. If it doesn't, then our receiver clock must be off. The receiver then applies clock corrections until the calculation of our position from each satellite lines up to the same place. A side benefit is that we now have a clock in our GPS receiver that is about as accurate as an expensive atomic clock.

Fig 24-9 shows an inexpensive handheld GPS receiver installed in my boat *Windfall*. In addition to

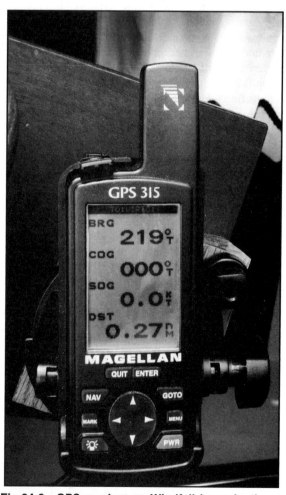

Fig 24-9—GPS receiver on *Windfall*, in navigation computer mode.

position data, it provides a navigation computer function as described for the LORAN receiver. Fig 24-9 shows this mode, providing information on how to get to the next waypoint, which is 0.27 nautical miles away, on a heading of 219° relative to True North. More advanced (and more expensive than $99) GPS receivers provide radar-like displays indicating the user's position on a map, for automobile use, or a chart for aviation or marine use. Manufacturers have digitized nautical and aeronautical charts so that they can be loaded into the receiver and provide a combined map and position display. See **Fig 24-10**.

As mentioned earlier, far more accurate readings can be obtained through auxiliary systems, such as *Differential* GPS, a system that requires a cooperating fixed ground station. If the ground station knows exactly where it is, and is relatively close to the mobile station, it can compare its known location with that predicted by GPS reception. Any difference must be due to the various (small) errors in satellite timekeeping or position recording since the signals pass through almost the same piece of space. The ground station can determine a correction for each received satellite at that moment and send it (on another frequency) to a DGPS receiver. The mobile GPS can apply the same corrections, with the resulting position determination accurate to within a few feet. The USCG has DGPS stations at critical locations, and private companies also provide such stations with a subscription fee, or for their own use as part of surveying or oil field exploration activities, for example.

GPS has found applications in many areas. Many long-haul truckers have GPS receivers connected to radio transmitters, so their progress and current position is instantly available to the company dispatcher. An Amateur Radio application, *Amateur Position Reporting System* (APRS) performs a similar function, with the results loaded onto the Internet, allowing anyone to keep track of an amateur's travel progress.

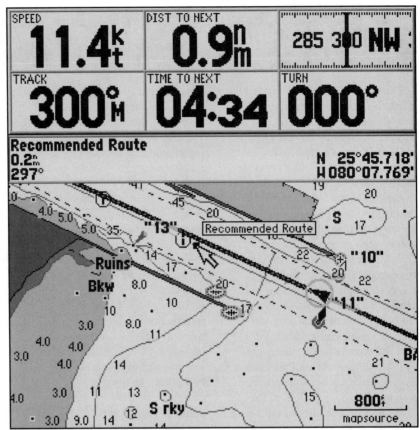

Fig 24-10—Display screen of The Garmin GPSMAP 172, a GPS receiver with integrated chart display and maritime navigation computer. (*Photo courtesy of Garmin International*)

Review Questions

1. To determine position using an RDF system, the navigator takes bearing measurements from his location to two known radio stations using the RDF. If those bearings are plotted on a chart, each line is a line of position (LOP) on which the navigator is located. The point at which the two LOPs cross then must be the location of the boat or aircraft—but is it? Suppose the navigator takes a third bearing to another transmitter. If everything works perfectly, it should cross at the same point, but it likely won't. Why?

2. What information can we infer from the difference in predicted positions of the third bearing with each of the other two in Question 1. above?

3. Besides VOR and ADF, another approach to air navigation is to have position and track determined by ground-based radar and communicated by radio from ground to the aircraft. Discuss the relative advantages and disadvantages of each approach.

Chapter 25

Where is Technology Taking Us Next?

How electronics have changed over time! At left, state-of-the-art, luggable handheld transceiver from the 1940s. At right, a 2005 bicycle equipped with several mobile VHF and UHF transceivers, a scanner, a timer/watch, digital compass, a portable weather station and an AM/FM entertainment console! Just in case, a Geiger counter is also available.

Contents

Is My Guess Better Than Yours?

That's a good question, and some of your guesses may be a lot better than mine! I guess I've had the opportunity to watch the evolution of a few generations of radio and other telecommunications systems, perhaps even to participate in the shaping of some directions. So I will pass along my thoughts and perhaps they will trigger different or better thoughts on your part.

What Can We Learn From The Past?

It has been said that "The Past is Prologue," implying that whatever has happened sets the stage for the future. That is certainly true, however, many read too much into trend data. The fact that a stock has gone up every Friday for the past five years doesn't mean anything—after all, your favorite football team has *always* beaten the point spread when playing at home, so surely they'll do so this next Friday... I have had friends lose lots of money on both kinds of bets!

What then are the trends that we can observe to see if they will have an impact on radios in the future? There are several that I think are key.

Processor Speed is Increasing

Computers are playing more and more significant roles in all electronics, including radio and other telecommunications systems. We have watched computers move from being extremely expensive, room-sized and unreliable monsters in the 1940s to inexpensive integrated electronic devices appearing in everything from toasters to palm-size general-purpose computers. It is interesting to observe the change in the processor clock frequency in typical home computers since they became widely available in the early 1980s.

Fig 25-1 shows how the clock speed has risen for the PCs (personal computers) over the past 25 years. Of course, my wife will point out that many of these computers are still in my basement, made obsolete by the next round of technology! The trend is clear—PCs no longer can limit your typing speed.

Storage and Processing Capability of Personal Computers

Computer RAM memory provides an general indication of how much data can be manipulated quickly.

Memory size has not been as fundamental a quantity as clock speed, since you can usually upgrade the amount of memory without changing the rest of the computer. The key ingredients here is the computer architecture, especially as it deals with memory:
- How much memory can be addressed by a particular processor design?
- Memory packaging—how much memory can you fit on a reasonably sized chip and what is the cost for such a chip?
- How much money does it make sense to put into memory? In other words would it be smarter to upgrade the memory or the whole computer?

So memory size is an interesting parameter to observe as a measure of computer "horsepower" and I've provided a chart in **Fig 25-2**.

Equipment Can be Made Smaller and Lighter

Each new generation results in equipment that is shrinking in size. This is a result of a couple of technological trends—the ability to make smaller and smaller, or increasingly complex, integrated circuits, and the higher processing speed of processors. The net result is

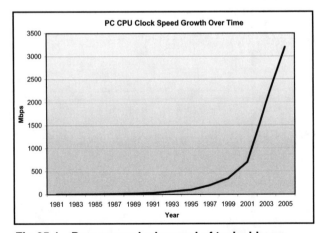

Fig 25-1—Processor clock speed of typical home PCs over the past few decades.

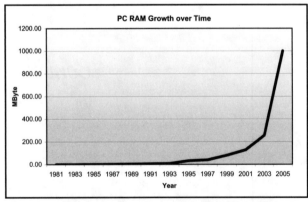

Fig 25-2—Random access memory (RAM) in typical home PCs over the past few decades.

that equipment is able to handle tasks once assigned to multiple pieces of gear with specialized functions.

So Where Does This Lead us?

How Small can you go?

While some parameters lead us to expect growth to double every few years, there are some fundamental size limits we come up against. It seems to me that the key "human factors" issues are:

• How small a display screen can you actually get useful information from?
• How many buttons do you need to send information, or to request different screens?

The above seem to me to set the limit on how small you can get based on current "man-machine-interface" (MMI) concepts. While a telecommunications device could conceivably be made to fit on a penny sometime in the near future, the risks of accidentally swallowing it while speaking into the microphone integrated into such a device might outweigh any benefits coming from a small size!

Someone might point out that a miniature communications device could become part of something else, perhaps a computer or a telephone, but in my opinion that just moves the problem around—it's now bigger because of what it's part of!

So What is Likely to Happen Next?

I think we're starting to see what the future holds as I write this. In 2005, manufacturers are faced with the dilemma of technology allowing devices to get smaller than can be gainfully employed and are responding with more functions in each device instead of making devices too small to be useful.

The Cell Phone as a Case Study

The result—cell phones that also serve as cameras and text messaging devices. The logical common denominator seems to be that since your cell phone needs a touch pad, a display and a battery why not make them a useable size for a person and design around the resulting box? What this means is that as technology gets smaller there will be more and more functions in a box that you have to carry anyway. Since the box itself, advertising and distribution tend to cost more than the electronics inside the box, the competitive marketplace

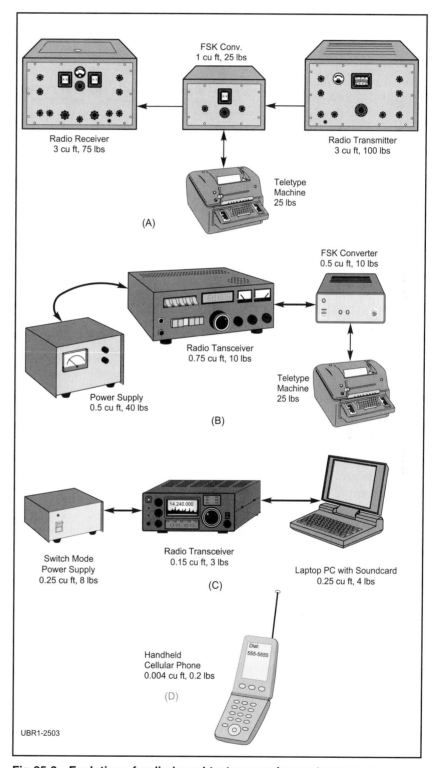

(A)

(B)

(C)

(D)

UBR1-2503

Fig 25-3—Evolution of radio-based text messaging systems as an example of the trend toward miniaturization. A-1940s, B-1970s, C-1990s, D-2005.

will tend to keep the price constant while the features expand.

So what might you expect to see included in the *Nextgen* cell phone? How about the following features all hanging from your belt?

- A GPS receiver, so your cell phone and your callers will know where you are, hopefully with an on/off switch if you want your location to be private.
- A navigation system like those in modern automobiles that use GPS and stored maps to get you where you want to go.
- A broadcast and satellite TV receiver. While stereo or five-channel reception might be possible, it would probably work better with headphones than with a bunch of little speakers in a small cell phone!
- A radio receiver capable of receiving frequencies almost from "dc to daylight."
- A full-function palm-top computer with the ability to infrared updates to and from your desktop or laptop computer.
- A dictation machine to allow you to record memos and letters while you're stuck in traffic.
- And, of course, a full-featured telephone responding to audio commands.
- Add your guesses here.

What About "Serious" Radio Equipment?

While the cell phone is a great example of emerging technology, with any luck there still will be other future devices larger and heavier than cell phones. Cell-phone technology makes good use of many repeater stations with power control that limits the output power of most phones to less than 1 W; there will always be a need for higher-powered radios capable of direct long-haul communication.

The current generation of HF transceivers is showing signs of following the same trend as the cell phones noted above. While some product lines are trending towards miniaturization, others seem to be going in the other direction. With

Fig 25-4 ICOM IC-7800 HF/VHF transceiver. The multifunction display can serve as a spectrum analyzer as shown here, along with the display of basic operational information, or can be set to decode radioteletype text.

larger radios, more and more functions that used to be outside the basic radio are being included as part of the package. In some new radios, not only is the FSK-to-data converter built in the radio, but also the data is decoded and displayed on a built-in general-purpose display. A good example of this trend is shown in the ICOM IC-7800 transceiver shown in **Fig 25-4**.

What Else is Happening?

Technology is moving forward on a number of fronts. One result of higher processor speed is the capability of digital signal processing (DSP) working at progressively higher speeds. The first DSP processors operated at audio frequencies. They did a great job of shaping audio bandwidth filters with very steep skirts that far surpassed the capabilities of analog filters. These filters could do a great job, provided that earlier stages of a receiver weren't overloaded or driven into distortion.

In the last few years, the higher speeds possible in CPU chips have resulted in special-purpose DSP chips able to move "closer to the antenna" of a receiver. This has meant that rather than just being able to shape the frequency response of the receiver audio stages, the current generation of DSP chips can move into the IF region of a receiver and eliminate out of band signals. This results in an architecture that can offer better overall receiver performance. The current crop of DSPs

operate at IF frequencies in the tens of kHz, generally requiring an extra conversion stage to get the IF down to the operating frequency of the DSP. As the speed of DSPs increase, they will be able to operate at IF frequencies in the MHz range, perhaps eliminating an extra conversion stage and all the distortion products that can occur there.

The trend is clear—as the speeds of DSPs increase, functions can move right up to the antenna terminals. This implies a complete multimode, all-frequency receiver on a single chip. Of course, we are faced with the question of where to put all the control knobs—one option is a large, mostly empty box with good controls and displays on the front panel—the other is a virtual receiver with controls and displays all on the screen of a personal computer. Both options are available as I write this and they will just continue to get smaller and work better.

And on the Transmit Side?

All of the benefits that DSPs bring to receivers are available in transmitters as well. DSPs can be used as audio processors to shape and optimize microphone and other input signal response, generate SSB signals, limit bandwidth, etc. As the processing speed increases, the functions again can move closer and closer to the antenna, or at least closer to the output power amplifier, which will likely not be on the same chip for some time—at least not for power outputs above a few watts.

How About the Far Side?[1]

Twenty some years ago, Swiss watchmakers made the classic mistake of trend watchers everywhere—they assumed that the trend lines would continue forever. Based on historical evidence (over 500 years of data), they concluded that they would continue to dominate the wristwatch business. They only had to make their finely crafted watches operate more accurately, be quieter and fit into progressively slimmer cases. It worked until someone (interestingly, also from Switzerland) developed the electronic watch. All of a sudden it no longer mattered that their watches were slim and quiet, the electronic watches could beat them on every parameter and sell for a small percentage of the price of a finely handcrafted Swiss watch.

The lesson is clear. Trend watching is useful, but you must also keep your eyes on the horizon. The catch phrase is "paradigm shift"—meaning looking at things in a completely different way. Another word for this is "thinking outside of the box"! I will go a bit out on a limb here, and it will be interesting to see if this is "old hat" by the time you read this! Imagine:

1. Phase 1—Devices are too small for useful input/output controls:
- Because devices are becoming too small for their controls, buttons will be replaced by an audio command interface; eg, *"Telephone, connect to Molly at the office."*
- Displays will suffer from the same constraint. Rather than force users to wear magnifying lenses to read the display, the display will appear in the lenses of the user much as aircraft "heads-up" displays seem to appear on the aircraft windscreen.

2. Phase 2—Bioengineering development continues at its current pace and the Man-Machine-Interface (MMI) moves from the edge of the machine to the edge of the user:
- Telecommunications devices become so small that they can be implanted *into* each user.
- The MMI moves into the user. This means that all control operations, input and output signals and displays occur directly inside the user's brain.
- Eventually, we could become part of how we communicate. We use our voice to talk to someone standing next to us—our internal communicator to talk to anyone else—without even thinking about it.
- With the processing built-in, we won't have to worry about directions. We just wonder about how to get somewhere and the built-in GPS and mapping functionality translate into directions just as we need them.
- We will be able to watch television just by thinking about the program—wherever we are. No more fighting over the remote—we all have one built in.

Well, that's my vision for what might be possible in the future. What's yours?

Notes

[1] With apologies to Gary Larson.

Appendix

Radio Construction Projects

Contents

One way to get a good feel for the principles outlined in this book is to build some of the circuits I described in the text. In this Appendix, I describe circuits that can be built successfully by those just learning about radio. By building the various circuit blocks in this section, you should be able to imagine more complex systems that are made up of combinations of these blocks.

Most of the circuits described here are not terribly fussy about construction techniques, or even the exact parts used. They can be built in a number of different ways, which I will describe. If there are any areas that need particular caution, I will point them out.

If you are not inclined to actually build these circuits, there is still something to be gained through a careful examination of the schematics and photos to visualize the transition from a *paper design* to an actual physical working model.

Getting the Parts

One of the major challenges of this kind of effort is finding the parts. There was a time when an electronic retailer seemed to be right on every street corner, providing all the electronic parts you could imagine. Those days are gone; however, there are still a number of dealers who can provide the parts you need by mail or through the Internet. It is likely that the availability of some parts might change over the life of this book, so be prepared to search on your own.

In general, I have tried to avoid specifying expensive parts. I don't recommend that you spend a lot of money to get the exact part that I may have identified. In many cases a slightly different value or a part with a different rating may be available at lower cost. Consider the resonant circuit designed to select the receiving frequency of the simple crystal set, or the transmitting frequency of

the simple transmitter. If the exact values for inductor and variable capacitor are not available, the tuning range for parts you do have may be different. The tuning range can be calculated from Eq 2-1 in Chapter 2.

If you are in a part of the world that has strong shortwave broadcast stations, you may even choose to redesign the tuned circuit to cover their frequencies. Feel free to experiment with different values of resistors and capacitors as well. Just be careful that you don't change the values of resistors setting the dc currents of the transistors by more than a factor of two to avoid transistor damage. For the transistor, almost any small-signal NPN transistor

should operate in place of the device specified. If you have a PNP transistor, the same circuit can work—just turn the battery connections around. If you wish to use a different audio integrated circuit for the audio stage, look for application notes on the Internet and compare the wiring and components with the one I used. If the performance is similar, it should work as well.

A good source of parts can be found in discarded electronic equipment. Careful use of wire cutters can yield a large number of potentially useful parts, especially if you have a meter that can read the resistance and capacitance of the

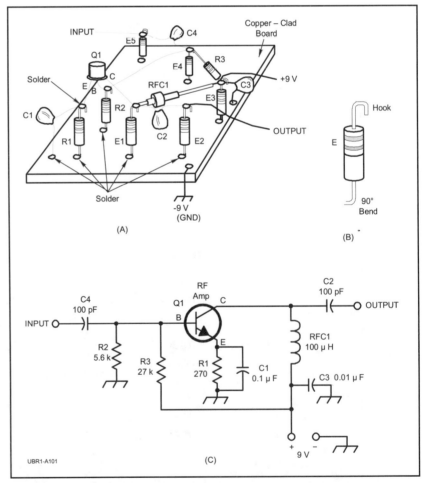

Fig A1-1—A sketch of a circuit and the corresponding layout using the ground-plane construction technique.

often obscurely marked parts. Just be sure that the equipment from which you harvest your parts will never need to be used again, because it certainly won't operate after this activity!

Construction Techniques

There are many ways that simple circuits can be constructed. I have chosen to use manufactured *perforated board* for my examples because it was available at the retailers I purchased the parts from. Perforated board is handy because the perforations can be used to easily locate the parts on one side and the wiring can be on the other. Depending on the materials available, you may wish to use other techniques.

- *Breadboard construction*—this may be the easiest to find, just a piece of wood and some nails will do the trick. Pick a dry piece of wood, draw your circuit on the board, put nails at the junction points and wrap the wires around the nails. Don't count on the nail for connection though, just for mechanical support. Solder the wires together at each junction. While this won't work well for some circuitry, it should work fine for these projects, and is perhaps the oldest construction technique.
- *Ground plane construction*—makes use of often readily available surplus unused printed circuit board material with a copper conductor on one or both sides. Use the copper foil as the common ground connection for the circuit by soldering the ground side of any grounded components to it. The other end of the components, plus any ungrounded ones, are elevated above the board by the height of the grounded components. **Fig A1-1** shows an example of how this technique is employed.

Component Handling

Some solid-state components are susceptible to damage from static electricity, also known as *electrostatic discharge* (ESD). A large static charge can punch through a semiconductor junction, rendering the part useless. Generally, once the part is

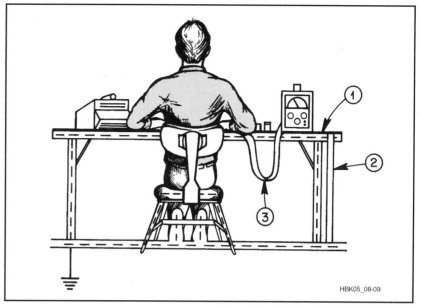

Fig A1-2—A sketch of a workstation designed for low risk of ESD damage. At (1) is a grounded dissipative mat, at (2) a grounded wrist strap connection and (3) a cable to the wrist strap.

HBK05_08-09

wired into the circuit, any such charge will be dissipated by the surrounding passive components and there will be less of a threat. Germanium devices, such as the 1N34A diode, are particularly susceptible. However, even modern silicon devices can be damaged.

The best way to avoid ESD component damage is to do all your work at work station that is especially set up to avoid ESD problems. Such a work station is shown in **Fig A1-2**. The key features are a grounded work surface, such as the special mat shown and the grounded wrist strap that keeps the technician's hands at the same ground potential as the bench. Any soldering equipment should also have the heated tip grounded to the same grounded work surface.

While such a work station is likely a good investment for people who are in the electronics business, the required investment is certainly not warranted (in my opinion) by concern over a $1.50 diode! Some reasonably easy and almost as safe methods can be employed, if appropriate care is taken.

- As you sit down to work, touch your hand to a grounded fixture, or even the outlet cover screw on a

grounded ac outlet box. Repeat every time you move your feet, and before you touch a sensitive device.
- Don't move your feet, so you don't create static charges.
- Pick up sensitive devices by one lead, using insulated tools if possible.
- Quickly wire into the circuit, one lead at a time.

Another concern about active devices is *thermal stress*. Too much soldering heat can also damage sensitive solid-state devices. As with ESD, there are a few simple rules that should help eliminate this problem:

- Place a heat sink—it can be as simple as an alligator clip—on each wire between the solder joint and the device before applying heat.
- If there are multiple connections to a solid state device, solder the device in place after all other connections are soldered together first.
- Use enough heat so that solder will flow quickly and remove the heat as soon as the solder flows.

Taking such precautions should eliminate most device failures from either ESD or too much heat.

A Simple Crystal Set

Yes, you can actually build our own crystal radio receiver. As noted in Fig 2-5 in Chapter 2, there aren't many parts required for a simple receiver. It may, however, be hard to find the high-impedance headphones that were popular when crystal sets were "high tech" back in the 1920s. The values for the other parts are not too critical, but should be fairly close to the values shown.

For the crystal we can avoid the difficulty of finding old-fashioned galena crystals and buy a very inexpensive modern semiconductor diode. It will work better and needs no adjustments. Almost any diode will work, but your radio will be more sensitive with a germanium diode, such as a 1N34A, than with a

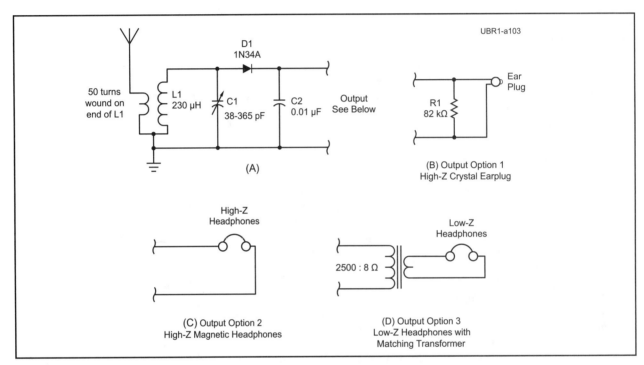

Fig A1-3—Circuit diagram (schematic) of a simple crystal set. Parts list for simple crystal receiver.

Part Number	Value	Type	Dealer	Dealer Part Number
C1	38-365 pF	variable capacitor	Antique ES[1]	C-V365
	20-220[2] pF (less tuning range)	variable capacitor	Philmore[3]	86-1450
C2	0.01 µF	ceramic capacitor	Mouser[4]	539-CK05103K
D1	1N34A	germanium diode	Antique ES	PQ972
L1	230 µH	ferrite rod antenna coil	Antique ES	P-C73
Phones	high-Z crystal earphone	Antique ES	P-A480	
R1	82 kΩ, 1/2 W	resistor[5]	Antique ES	R-1820k
		Mouser	660-CF1/2L823J	

One of the following is needed if using low-Z headphone with this receiver. See also Project 2.

T1	2500[6]: 8 Ω	output transformer	Antique ES	P-T983
T1 (less expensive)	1000: 8 Ω	output transformer	Radio Shack[7]	273-1380

Mechanical and miscellaneous parts suggested for this and subsequent projects

Type	Dealer	Dealer Part Number
Hook-up wire 90 feet # 22 AWG solid copper	Radio Shack	278-1221
Heavy duty 9 V battery snap connector	Radio Shack	270-324

more common silicon diode, such as a 1N914.

Will it Actually Work?

How well you can receive stations with this simple receiver will depend on how close you are to an AM transmitter, the power of that transmitter and the length and the effectiveness of your antenna and ground connections.

The most effective antenna you can probably manage for this simple radio is a piece of wire as long and as high as practical. In theory a vertical wire that $1/4$ wavelength long would be optimal. This makes up half of a *dipole antenna* and requires a good ground for the other side of the connection. Look at your answers to Review Question 2 in Chapter 1 and divide by two to compute the length of a $1/4$-wave antenna.

But this is probably still longer than you can deal with! So your antenna will be shorter than optimum, and will therefore pick up less signal. As a young lad, I had good results with about 30 feet of wire run to a tree outside my second floor bedroom window, perhaps 10 to 15 feet above ground. I was able to hear strong broadcast stations from about 25 miles away.

For a ground connection, the classic connection point in a house is a cold-water pipe. This works because pipes are often copper and they end up underground outside the house. Make sure there's no PVC (plastic) pipe in the path, especially if your house was built fairly recently. Plastic pipes do not conduct electricity! Another ground possibility is to connect to the screw on the outside of a power outlet plate. If proper grounding has been employed in your house's wiring, this should connect all the way back to the ac mains service-entrance ground.

Putting it Together

The schematic of the crystal receiver is shown in **Fig A1-3**. I placed the parts for this project on one end of the perforated board to leave room for the RF amplifier circuit coming in Project 4. Build the circuit taking into account the cautions noted in the introduction above. Otherwise there should be few restrictions on technique. The hook-up wire I specified in the parts list comes in three rolls, one each red, black and green. While it isn't necessary to use colored wire, I propose that if you do you use the following coding:

- *Red*—Use for positive power supply connections.
- *Black*—Use only for connections at ground potential.
- *Green*—Reserve all other connections.

Options

Note that there are a few options for connection of headphones for this receiver. In the days of crystal sets, headphones were designed for optimum performance when used with such a receiver. They were constructed with many turns of magnet wire to provide a high impedance load, to avoid loading down the circuit. Impedances of 2000 Ω and higher were common.

If you can find such headphones, by all means connect them directly to the output. You will have the best results that can be obtained. If you can't, all is not lost. There are three other possibilities, described below:

- *Use a high-impedance crystal earplug*—note that this is not the same kind of crystal as our detector. Rather this is a Rochelle salt crystal *piezoelectric* device that converts electrical signals into vibrations. It does not have dc continuity, so the resistor shown in Fig A1-3 is needed to complete the circuit for dc. Note that many such earplugs are low-impedance devices and will not work with this receiver. So be sure of what you are getting!
- *Use low-impedance earphones with a matching transformer*—an audio matching transformer can be used to transform the low impedance of modern stereo or other headphones to a higher impedance to use with the crystal set. If you are using stereo headphones, you can either use one channel or tie the two in parallel. Note that the resistor in Fig A1-3B is not needed in this configuration, since a dc path is provided through the transformer primary.
- *Wait for Project 2*—One option in Project 2 will provide a low-impedance output. If it works off the bat you will be fine. If not, by combining Projects 1 and 2 without testing Project 1 first, it may become more difficult to troubleshoot a problem, if you have one.

The basic crystal set can work successfully with strong signals if a germanium 1N34A diode is used. The more common silicon diodes

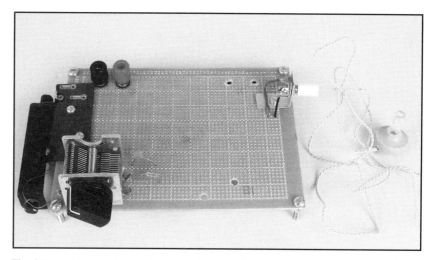

Fig A1-4—Photo of top of completed crystal receiver using perforated board construction.

will work, but require a very strong signal to be detected due to the knee of their operating curve starting at about 0.6 V, compared to around 0.1 V for a germanium diode.

Fig A1-4 and **Fig A1-5** are photographs of the receiver built by author W1ZR.

Notes

[1] Antique Electronic Supply (Antique ES) can be found at **www.tubesandmore.com** or 480.820.5411

[2] This part will work, but will have a limited tuning range.

[3] Philmore products are available at some electronic retailers.

[4] Mouser Electronics can be found at **www.mouser.com**.

[5] Not needed if using high-Z magnetic (not crystal) headphone.

[6] A higher value primary impedance would provide increased sensitivity. Values as high as 10 kW were common in vacuum tube radios and are sometimes found as discards.

[7] Radio Shack can be found at **www.radioshack.com**.

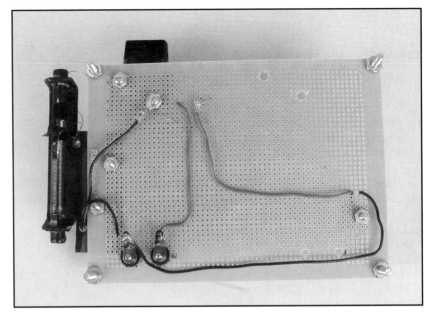

Fig A1-5—Photo of bottom of completed crystal receiver.

Project 2

Adding an Audio Amplifier to the Simple Crystal Set

As we noted in Chapter 3, a major improvement in the usefulness of a crystal set came with the addition of a vacuum-tube audio amplifier, a technology that was developed during WW I. For this project, I will skip over vacuum-tube technology and add a simple single-stage transistor amplifier to the simple crystal set. This amplifier, shown schematically in **Fig A1-6**, will amplify all the signals that come from the crystal set—whether you want to hear them or not, of course!

While this amplifier could easily be built on the same board as the crystal receiver, I chose to build it on a separate board that will later be paired with the integrated circuit power amplifier in Project 4. That combination can be used for other audio projects.

As shown in Fig A1-6, the transistor amplifier can be configured to drive either a high- or a low-impedance load. If you don't have high-impedance earphones, it may be less expensive to build this simple amplifier rather than trying to locate a transformer or another set of headphones.

The high-Z-in to high-Z-out transistor amplifier is called a *common emitter amplifier* and it

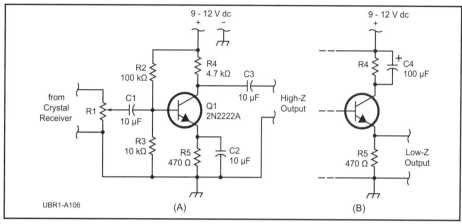

Fig A1-6—Circuit diagram of the transistor audio amplifier for the simple crystal set. Parts list for transistor audio amplifier add-on to crystal receiver.

Part #	Value	Type	Dealer	Dealer Part Number
C1,2,3	10 µF, 25 V	electrolytic capacitor	Mouser	140-HTRL25V10
C4	100 µF, 25 V	electrolytic capacitor	Mouser	140-HTRL25V100
Q1	2N2222	silicon NPN transistor	Mouser	610-2N2222A
R1	10 kΩ, 1/2 W	potentiometer	Antique ES	R-VA10KL
R2	100 kΩ, 1/2 W	resistor	Mouser	660-CF1/2L104J
R3	10 kΩ, 1/2 W	resistor	Mouser	660-CF1/2L103J
R4	4.7 kΩ, 1/2 W	resistor	Mouser	660-CF1/2L472J
R5	470 Ω, 1/2 W	resistor	Mouser	660-CF1/2L471J

provides a voltage gain. The high-Z-in to low-Z-out configuration is called a *common collector amplifier* and provides a power or current gain, but no voltage gain. That sounds a lot like a combined power amplifier and impedance transformer—just what we want if we have low-Z phones!

The VOLUME control at the input may reflect a bit of optimism on my part. If the potentiometer is not available, a resistor of 10 to 100 kΩ could be substituted with C1 con-

nected to the top end. Just be ready to pull the phones off if you encounter a very loud station!

The supply voltage shown is 12 V; however, if a bench type supply is not available the amplifier can be powered by a 9-V "transistor-radio" battery in the holder listed previously. If desired, a small switch can be inserted in the line from the battery, or you can remove the battery when it is not being used. See **Fig 1A-7** and **Fig 1A-8** for photos of my completed amplifier.

Fig A1-7—Photo of top of completed transistor audio amplifier.

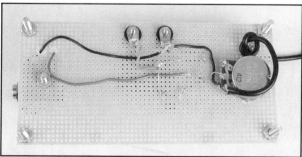

Fig A1-8—Photo of bottom of completed transistor audio amplifier.

Adding an RF Amplifier Ahead of the Simple Crystal Set

The preceding audio amplifier increases the level of signals, interference and noise coming from our crystal set. This makes it easier to hear a station we are tuned to, but also amplifies all the other limitations of our simple receiver. As noted in Chapter 3, another way to improve the performance is to add a single-stage transistor RF amplifier ahead of the simple crystal set.

The RF amplifier in **Fig A1-9** increases the level of signals reaching the detector, rather than the audio signals leaving it. This increases signals reaching the detector so that it operates in the more linear portion of its response curve. Perhaps more importantly, you have added an additional tuned circuit to improve the receiver's *selectivity*, the ability to separate stations. This second tuned circuit is likely to make much more of an improvement than the first since the selectivity of the first tuned

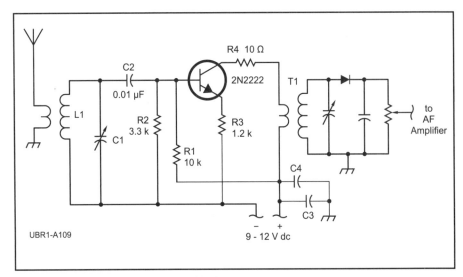

Fig A1-9—Circuit diagram of the transistor RF amplifier for the simple crystal set.
Parts list for transistor RF amplifier add-on to crystal receiver.

Part #	Value	Type	Dealer	Dealer Part #
C1	38-365 pF	variable capacitor	Antique ES	C-V365
	20-220[1] pF	variable capacitor	Philmore	86-1450
C2, 3	.01 µF, 25 V	ceramic capacitor	Mouser	539-CK05103K
C4	10 µF, 25 V	electrolytic capacitor	Mouser	140HTRL25V10
Q1	2N2222	silicon NPN transistor	Mouser	610-2N2222A
R1	10 kΩ, 1/2 W	resistor	Mouser	660-CF1/2L103J
R2	3.3 kΩ, 1/2 W	resistor	Mouser	660-CF1/2L332J
R3	1.2 kΩ, 1/2 W	resistor	Mouser	660-CF1/2L122J
R4	10 Ω, 1/2 W	resistor	Mouser	660-CF1/2L103J
R5	100 Ω, 1/2 W	resistor	Mouser	660-CF1/2L101J
T1	230 µH	antenna coil	Antique ES	P-C70

Fig A1-10—Photo of top of crystal receiver with RF amplifier.

Fig A1-11—Photo of bottom of crystal receiver with RF amplifier.

circuit was limited by the unavoidable loading of the connected antenna.

With the combination of a tuned RF amplifier ahead of the detector and an AF amplifier following it, we have the modern solid-state equivalent of the Atwater-Kent TRF (tuned radio frequency) receiver as shown in Chapter 3. The Atwater-Kent had three tuned circuits, and associated variable capacitors, while ours has two, but the operation is similar. If desired, a second tuned RF amplifier could be inserted between the first RF and the detector and ours would become a true "Three-dial TRF!" The more gain we insert with tuned stages, the more we must worry about inadvertent feedback and having the RF amplifier turn into an oscillator. We intend to investigate oscillators later, so a single stage that doesn't oscillate inadvertently may be a more satisfying exercise.

To minimize feedback, the RF amplifier inductor should be installed in a shielded enclosure. This should be tied to the common ground through as short a lead as possible, and all other leads should as short and direct as possible. In the Atwater-Kent design, the RF inductors were oriented so that they were mutually perpendicular to minimize magnetic coupling. In addition, the larger size of that receiver resulted in the inductors being far apart.

In this design, I have intentionally reduced the gain somewhat to improve the likelihood of success; however, if there are any problems with oscillation as the two tuned circuits are tuned to the same frequency, reorienting the inductors may help minimize the problem. If there is no sign of instability, the value of R3 can be reduced to an ion as 100 W to increase the gain. The photos in **Fig 1A-10** and **Fig 1A-11** show the completed RF and AF amplifiers with the crystal receiver.

Adding an AF Power Amplifier to Drive a Loudspeaker

So far, your efforts have been oriented towards improving the crystal receiver to optimize performance for a single listener using earphones. Since you now have a receiver "ready for prime time," you may want to add more AF amplification to drive a loudspeaker so that others can hear what you've made. To add some variety to these projects, instead of discrete transistors and associated components, I have elected to use an integrated circuit (IC) in this stage, as shown in **Fig A1-12**.

I have shown the new power amplifier stage together with the earlier AF amplifier. The earlier amplifier now becomes a *preamplifier* for the output IC amplifier. While each stage could be built separately and powered from individual batteries, I will use a common power source here. Good practice requires that the power feeds to the stages be *decoupled* so that signals will not inadvertently couple between the stages, causing instability. The added decoupling is shown in Fig A1-12.

Note that by going to an IC amplifier, you have eliminated the need for bias setting resistors, as well as some other components. You also gain some design flexibility that I have elected not to take advantage of in the interests of simplicity and

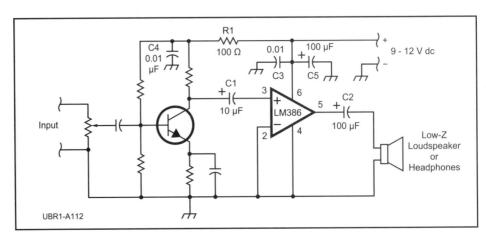

Fig A1-12—Circuit diagram of the IC AF speaker amplifier with preamplifier.
Additional parts list for integrated circuit audio amplifier add-on to crystal receiver.

Part #	Value	Type	Dealer	Dealer Part #
C1	10 µF, 25 V	electrolytic capacitor	Mouser	140-HTRL25V10
C2,5	100 µF, 25 V	electrolytic capacitor	Mouser	140-HTRL25V100
C3,4	.01 µF, 25 V	ceramic capacitor	Mouser	539-CK05103K
IC1	LM386N	integrated audio amplifier	Mouser	513-NJM386D
IC socket (8-pin dual in-line)			Mouser	517-ICA-083-WBTG30
Grid type IC spaced PC board 2³/₄" by 6"			Radio Shack	276-1395
R1	100 Ω, ¹/₂ W	resistor	Mouser	660-CF1/2L101J

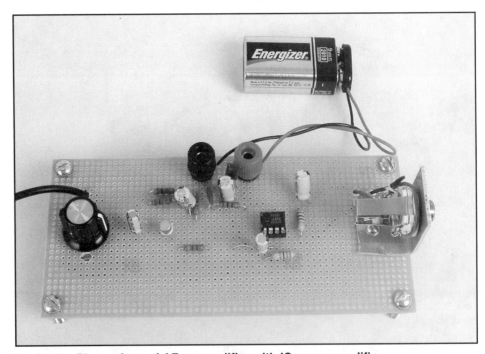

Fig A1-13—Photo of top of AF preamplifier with IC power amplifier.

to maintain focus on the *radio* aspects of this project. If you are interested in finding out more, you can find a data sheet for the output IC on the Internet.

One downside of working with ICs is that the connections are quite close together, making interconnections a bit trickier. I have specified an IC socket on the parts list to provide a bit more space for connections. **Fig A1-13** and **Fig A1-14** are photos of the completed IC output amplifier with AF preamplifier.

While the power amplifier and preamps are shown powered by a 9V transistor battery, the drain is much higher than for either of the earlier projects resulting in a short battery life. This would be a good project to run from a 12 V power supply.

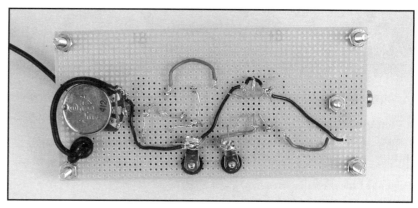

Fig A1-14—Photo of bottom of AF preamplifier with IC power amplifier.

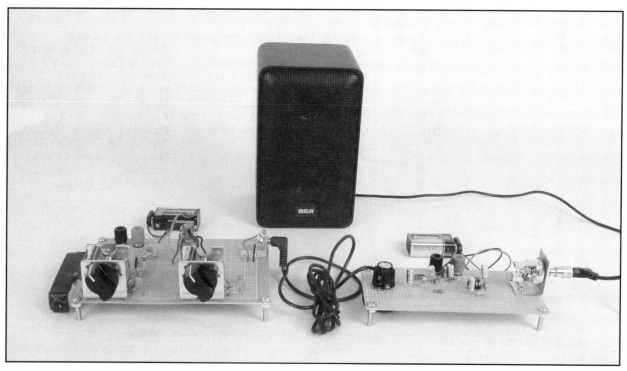

Fig A1-15—Photo the complete TRF receiving system. On the left is the RF amplifier and crystal detector; on the right is the AF preamplifier and power amplifier. In the rear is a small loudspeaker.

Project 5

A Simple AM Broadcast Band Radiotelegraph "Transmitter"

Before you proceed with this project, it's important to note that anytime you get into the realm of "radio transmitters" you need to be sure of local regulations. In the US, equipment of this sort can only be used if it does not cause *harmful interference* to others. Since a broadcast band transmitter, by definition, has the potential to interfere with those making lawful use of licensed services, it is the responsibility of users to avoid such interference. A key piece of the FCC rules is provided below:

Title 47—Telecommunication Chapter I—Federal Communications Commission; Part 15—Radio Frequency Devices; Sec. 15.5 General Conditions of Operation:

a. *Persons operating intentional or unintentional radiators shall not be deemed to have any vested or recognizable right to continued use of any given frequency by virtue of prior registration or certification of equipment, or, for power line carrier systems, on the basis of prior notification of use pursuant to Sec. 90.63(g) of this chapter.*

b. *Operation of an intentional, unintentional, or incidental radiator is subject to the conditions that no harmful interference is caused and that interference must be accepted that may be caused by the operation of an authorized radio station, by another intentional or unintentional radiator, by industrial, scientific and medical (ISM) equipment, or by an incidental radiator.*

c. *The operator of a radio frequency device shall be*

required to cease operating the device upon notification by a Commission representative that the device is causing harmful interference. Operation shall not resume until the condition causing the harmful interference has been corrected.

What this means is that any experimentation with a transmitter such as is described below should take place with no antenna, or the minimum antenna necessary to transmit usable signals within your residence. Connecting to an outside antenna could place you in serious jeopardy of violating federal law.

Fortunately, if you wish to pursue such activity, the FCC, and regulatory agencies of almost all countries,

support the Amateur Radio Service—a service designed for exactly such experimentation. See Project 6 for a small, but serious, HF Amateur-Radio transmitter capable of worldwide communications under optimum conditions.

Those in jurisdictions outside the US should check with appropriate regulatory agencies to make sure such devices are legal, or find out if any specific restrictions on operation apply.

The Simplest of Transmitters

For the example of a transmitter I have selected a single-transistor oscillator circuit (**Fig A1-16**) that is easy to build. In a real-world transmitter, such an oscillator would be followed by multiple RF amplifier

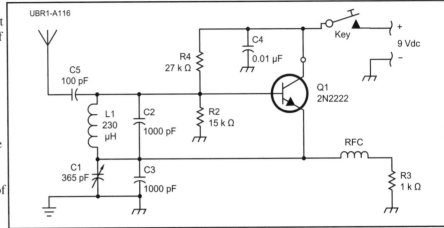

Fig A1-16—Circuit diagram of the simple AM broadcast-band radiotelegraph transmitter.

Parts list for simple AM broadcast band radiotelegraph "transmitter."

Part #	Value	Type	Dealer	Dealer Part #
C1	38-365 pF	variable capacitor	Antique ES[1]	C-V365
	20-220[2] pF	variable capacitor	Philmore[3]	86-1450
C2,3	.001 µF, 25 V	ceramic capacitor	Mouser[4]	539-CK05101K
C4	.01 µF, 25 V	ceramic capacitor	Mouser	539-CK05103K
L1	230 µH	ferrite rod antenna coil	Antique ES	P-C73
Q1	2N2222	silicon NPN transistor	Mouser	610-2N2222A
R1	27 kΩ, 1/2 W	resistor	Antique ES	660-CF1/2L273J
R2	15 kΩ, 1/2 W	resistor	Mouser	660-CF1/2L153J
R3	1 kΩ, 1/2 W	resistor	Mouser	660-CF1/2L102J
RFC	1.2 mH	RF choke	Antique ES	P-C206

stages. This is not too different from that of Project 3, raising the transmitter output power level to that needed to get the signal to your desired audience. In this project, the audience is minimal, perhaps just across the room!

To receive radiotelegraph signals in a receiver, you need a beat oscillator, as described in Chapter 9. Such a beat oscillator for our crystal receiver could actually be another transmitter adjusted in frequency to be about 700 Hz above or below the main transmitter frequency. An even simpler approach is to use a "non-cooperating" transmitter—a station that you can hear with a crystal set. By tuning your transmitter to a slightly different frequency above or below the radio station's frequency, you should hear an audible beat note between the two signals in your receiver. If you key the transmitter on and off, the beat note will go on and off and you can hear a simulated radiotelegraph signal.

I have not specified an actual telegraph key in the parts list since not everyone is sufficiently interested in Morse Code to want to buy one. You can just touch the battery supply wire on and off to give you the idea. Anyone who does want to pursue code practice could use such an oscillator as a starting point and borrow a key from almost any licensed Amateur Radio operator. Telegraph keys can be purchased from most Amateur Radio dealers, or a wide variety is available on the Internet, for example at Morse Express, **www.morsex.com**, starting as low as $11 (in 2005) for a useable beginner's practice key. I won't mention how pricey the hand-made versions can get!

Notes:

[1]**www.tubesandmore.com**

[2]This part will work, but with a limited tuning range.

[3]Philmore products are available at some electronic retailers.

[4]Mouser Electronics: **www.mouser.com**

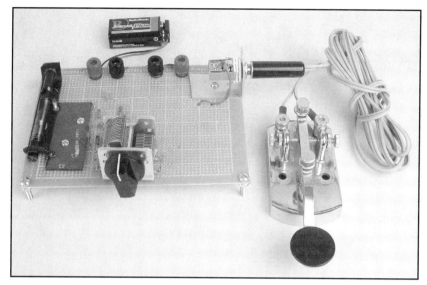

Fig A1-17—Photo of top of simple AM broadcast band radiotelegraph transmitter.

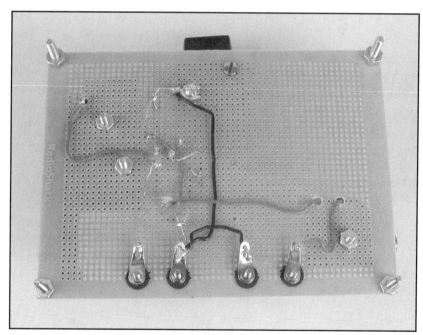

Fig A1-18—Photo of bottom of simple AM broadcast band radiotelegraph transmitter.

A Serious Low-Power Amateur Radiotelegraph Transmitter

Readers with an Amateur Radio operator's license—or those who intend to obtain one—may wish to construct an actual low power (QRP) transmitter for use on the 40-meter amateur band. "QRP" operation (generally considered transmitting at levels below 5 W output—this transmitter puts out only about 0.5 W) is a popular sub hobby in Amateur Radio.

A warning: It can be frustrating trying to be heard through noise and interference on 40 meters when you're using a "pip-squeak" transmitter. I don't recommend that a newcomer to Amateur Radio start out with really low power for his/her first transmitter. Most commercial transmitters are 100-W units, for example and you have a lot more chance to be heard at this level than you do at $\frac{1}{2}$ W!

Still, this little "Tuna Tin 2" transmitter is a fun and easy-to-build project that can work well, especially if used with a good antenna system. The following is a reprint of an article in the March 2000 issue of *QST*, the official journal of the American Radio Relay League.

If you want a serious operational challenge, the preceding audio amplifier, or a similar one, could be inserted at measurement point 2 in the diagram. The amplifier becomes an AM modulator. Change the crystal to 7290 or 7295, the usual AM operating frequencies and see if you reach anyone. However, radiotelegraph operation is more successful at these low power levels than AM operation.

The Tuna Tin 2 Today

Ham radio lost its kick? Go QRP with this weekend project! Worked All States with a 40-meter half-watter?
You betcha!

In the 1970s, the late Doug DeMaw, W1CER/W1FB, ARRL Technical Editor, was one of several Headquarters staff who published homebrew projects, many with a QRP twist. One of those was a simple, two-transistor 40-meter transmitter that used a tuna can as the chassis. Dubbed the "Tuna Tin 2," it was a popular project, introducing many hams to homebrewing and QRP. A series of events, some quite amazing, have come together to keep the magic alive—the original Tuna Tin 2, built in the ARRL Lab, is still on the air and articles, Web pages and kits are available for this famous rig. Some have dubbed the Tuna Tin 2 revival as "Tuna Tin 2 mania"—an apt term to describe the fun that people are still having with this simple little weekend project.

This article has been edited from the original, written by DeMaw and published in the May 1976 QST. You can download a copy in Adobe PDF format from the ARRL Members-Only Web site at: **http://www.arrl.org/members-only/extra/features/1999/0615/1/tt2.pdf**. Some of the original parts are no longer available, so modern components have been substituted, using values that were featured in a column in QRP with W6TOY on the ARRL Web Extra. I think that Doug would have been pleased to see just how popular that little rig still is, almost a quarter century after he first designed it and built it in the ARRL Lab.— *Ed Hare, W1RFI, ARRL Laboratory Supervisor*

The original Tuna Tin 2

Workshop weekenders, take heart. Not all building projects are complex, time consuming and costly. The TunaTin 2 is meant as a short-term, gotogether-easy assembly for the ham with a yen to tinker. Inspiration for this item came during a food shopping assignment. While staring at all of the metal food containers, recollections of those days when amateurs prided themselves for utilizing cake and bread tins as chassis came to the fore. Lots of good equipment was built on make-do foundations, and it didn't look ugly. But during recent years a trend has developed toward commercial gear with its status appeal, and the workshop activities of many have become the lesser part of amateur radio. While the 1-kW rigs keep the watt-hour meters recording at high speed, the soldering irons grow colder and more corroded.

A tuna fish can for a chassis? Why not? After a few hours of construction, 350 milliwatts of RF were being directed toward the antenna, and QSOs were taking place.

Maybe you've developed a jaded appetite for operating (but not for tuna). The workshop offers a trail to adventure and achievement, and perhaps that's the elixir you've been needing. Well, Merlin the Magician and Charlie the Tuna would probably commend you if they could, for they'd know you were back to the part of amateur radio that once this whole game was about—creativity and learning!

Parts Rundown

Of course, a tunafish can is not essential as a foundation unit for this QRP rig. Any 6½-ounce food container will be okay. For that matter, a sardine can may be used by those who prefer a rectangular format. Anyone for a Sardine-2? Or, how about a "Pineapple Pair?" Most 6½-ounce cans measure 3¼ inches in OD, so that's the mark to shoot for. Be sure to eat, or at least remove the contents before starting your project!

Although the original project used all RadioShack parts, some of the parts are no longer stocked. The 2N2222A transistor is

Kits and Boards

While the original Tuna Tin 2 can be built from scratch, surprisingly, printed-circuit boards and kits are still available.

The September 16, 1999 *QRP with W6TOY* column in the *ARRLWeb Extra* featured a modern version of the Tuna Tin 2[1]. FAR Circuits can supply the printed circuit for W6TOY's version (not built on a tuna tin) as well as the original design PC board[2].

JOE BOTTIGLIERI, AA1GW

W6TOY's version of the Tuna Tin 2 design—without the tuna can.

Those who want to buy everything all in one place can buy a complete kit, including PC board from the NJ-QRP Club[3]. Send a check for $12 postpaid to George Heron, N2APB, New Jersey QRP Club, 2419 Feather Mae Ct, Forest Hill, MD 21050. Doug Hendricks, KI6DS also designed a version of the Tuna Tin 2, for the Northern California QRP Club (NorCal)[4].

[1]See: **http://www.arrl.org/members-only/extra/features/1999/09/16/1/**.
[2] FAR Circuits, 18N640 Field Ct, Dundee, IL 60118-9269, tel 847-836-9148; **http://www.cl.ais.net/farcir/**
[3]NJ-QRP Club, contact: George Heron, N2APB, 2419 Feather Mae Ct, Forest Hill, MD 21050; **n2apb@amsat.org; http://www.njqrp.org/**. NJ-QRP has a section of their Web site devoted to the Tuna Tin 2 revival. See **http://www.njqrp.org/tuna/tuna.html**.
[4]Northern California-QRP Club (NorCal), 3241 Eastwood Rd, Sacramento, CA 95821; tel 916-487-3580; **jparker@fix.net; http://www.fix.net/NorCal.html**. Like the NJ-QRP Club, NorCal also has a Tuna Tin 2 revival page at: **http://www.fix.net/~jparker/norcal/tunatin2/tunatin.htm**.

widely available. The original coils have been replaced with inductors wound on toroidal cores. Printed circuit boards are available from several sources and the NJ QRP Club is offering a complete kit of parts. (See the sidebar "Kits and Boards".)

The tiny send-receive toggle switch is a mite expensive. The builder may want to substitute a low-cost miniature slide switch in its place. A small bag of phono jacks was purchased also, as those connectors are entirely adequate for low-power RF work.

Finding a crystal socket may be a minor problem, although many of the companies that sell crystals can also supply sockets (you can locate a number of crystal manufacturers and distributors on the ARRL TISFIND database at **http://www.arrl.org/tis/tisfind.html**). Fundamental crystals are used in the transmitter, cut for a 30-pF load capacitance. Surplus FT-243 crystals will work fine, too, provided the appropriate socket is used. If only one operating frequency will be used, the crystal can be soldered to the circuit board permanently. Estimated maximum cost for this project, exclusive of the crystal, power supply and tunafish, is under $20. The cost estimate is based on brand new components throughout, inclusive of the

---

The Tuna Tin 2 on the Road

Those who've read our on-line publication, the *ARRLWeb Extra*, probably saw the article that appeared in the June 15th edition titled "The Tuna Tin 2 Revival." This article told an incredible tale of how the original Tuna Tin 2 was lost from the ARRL Lab and was found years later in a box of junk under a fleamarket table in Boxboro, Massachusetts. The Tuna Tin 2 was refurbished by Bruce Muscolino, W6TOY, and put back on the air by me on June 4, 1999. Since that time, over 400 hams have had the pleasure of working the original Tuna Tin 2, some using their own Tuna Tin 2 rigs built in the 70s (or built anew from the available kits).

California Dreamin'

After making about a hundred contacts from home, I was asked to attend an IEEE meeting in Long Beach, California. My sister, Bev, lives in the area, so I planned a week-long visit. I tossed the Tuna Tin 2 and a G5RV into my suitcase, hoping to give a few West Coast hams a chance to make a contact with the original.

After all the hugs and kisses, I explained to my sister what I was up to. She grinned, remembering the wild days of my youth, climbing trees to string wires all over our property, back when I was WN1CYF. As I looked over the site, though, I was not too hopeful; about the best I thought I could do would be to try a random wire around the balcony, maybe risking a run over to a small tree or two. I looked roofward and sighed, "Gee, it would be nice to get an antenna up on the roof." She made a quick call to Debbie, the building manager and close friend, who winced painfully and said, "Don't fall off!" and, in a classic Schultz accent, "I know nothing!"

We took the G5RV up to the emergency roof access, walked boldly out, and I proceeded to string the antenna up while Bev stood guard. I got the antenna up, dropped the feedline past the upstairs apartment balcony and hoped for the best.

Sure enough, the "antenna police" were on alert—the tenant right below us heard the noise and wondered what was going on. Just as we got back to the apartment, the phone rang; it was Debbie. She told us of the complaint, told us the excuse she gave and wished us luck.

With Bev watching with great interest, I hooked up the Heath HW-8 I used as a receiver, hooked up the Tuna Tin 2, the code key and

antenna tuner, and gave the band a fast listen. Signals were booming in. On June 19, I worked my first contact with the Tuna Tin 2 from the West Coast, W6PRL/QRP. Every evening, after a day of offshore fishing, Bev and I expected to find that the antenna police had confiscated the wire, but somehow, it stayed up the whole week. By the end of the week, 45 new stations were in the Tuna Tin 2 log!

Among the Monsoons and ScQRPions

I was then asked if I would be willing to attend the ARRL Arizona State Convention at Ft Tuthill. That is an annual pilgrimage for many a QRPer; how lucky could I get? I agreed, but warned the ARRL Division Director that I might spend a bit more time away from the ARRL booth than usual. In the meantime, I casually asked Joe Carcia, NJ1Q, the W1AW station manager, if he could arrange for W1AW/7/QRP to be used at the convention. After some consultation with Dave Sumner, a new QRP "first" was in the works. In the meantime, the Arizona ScQRPions[1], an Arizona QRP club, asked me if I would give a presentation at the QRP forum they sponsor at Ft Tuthill every year. I agreed, but with one condition—they had to be willing to host W1AW/7/QRP at their booth. I would have loved to be a fly on the wall as that e-mail was read!

A *great* time was had by all, but W1AW/7/QRP did not go off without a hitch. An operator error (mine) damaged the receiver (the binaural receiver, designed by Rick Campbell). The local QRPers came through, though, and several receivers were made available to the operation to finish the day. Even worse, later in the day, it looked like all was lost! During a quick test of the Tuna Tin 2, one of the resistors emitted a puff of smoke, and the power went to 0 W. I had just blown up the original Tuna Tin 2!

I did a quick troubleshooting job and identified that the output transistor had short-circuited. Special thanks go to Niel Skousen, WA7SSA, who dug into his portable junkbox. (Niel is a real ham's ham! How many hams do you know who bring their junkbox to a hamfest?) He quickly located a 2N2222A. I handed him the Tuna Tin 2 and asked him if he would mind installing it. After that W1AW/7/QRP was back on the air.

After the convention, using a borrowed receiver, I took the Tuna Tin 2 on a whirlwind tour of Arizona, although I only got to operate two

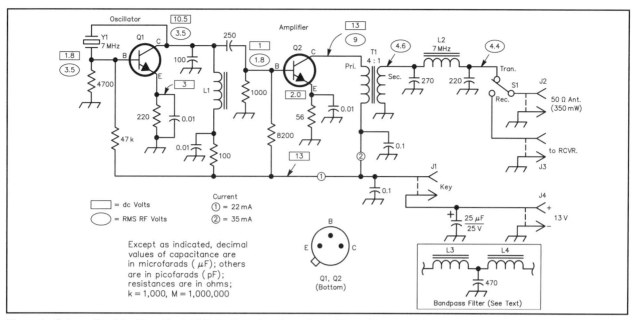

Figure 1—Schematic of the Tuna Tin 2 QRP rig. Note that the polarized capacitor shown in the schematic is an electrolytic.

J1—Single-hole-mount phono jack. Must be insulated from ground. Mounts on printed circuit board.

J2, J3, J4—Single-hole-mount phono jack. Mount on tuna tin chassis.

L1—22 μH molded inductor

L2—19 turns of #26 wire on a T-37-2 toroidal core

L3, L4—21 turns of #24 wire on a T-37-6 toroidal core

Q1, Q2—2N2222A or equivalent NPN transistor.

S1—Antenna changeover switch. Miniature SPDT toggle (see text).

T1—4:1 broadband transformer. 16 turns of #26 wire on the primary, 8 turns of #26 wire on the secondary, on an FT-37-43 toroidal core.

Y1—Fundamental crystal, 7 MHz.

nights from a campsite in Williams. I had brought along my DK9SQ[2] 33-foot portable fiberglass mast, so my antenna went up and down quickly. (Let me tell you, this is one great product. I literally put up my 40 meter inverted **V** in 5 minutes, 33 feet in the air. Taking it down was even faster.) It was monsoon season in Arizona and it rained each night. Despite the downpours, I doggedly squeezed in operating time in between thunderstorms, and added a few new ones to the log.

Hanging Out in the Park

Just two weeks later, I was off to Golden, Colorado for the Colorado State Convention (during which I got to show off the Tuna Tin 2 to the Colorado QRP Club[3]) and the trusty Tuna Tin 2 and portable mast came along with me. I scoped out the hotel area—no good. The noise level from the high-tension lines was just too high. The convention was held in a small park, so after the confab ended I walked a mile back to the hotel, loaded up the Tuna Tin 2, batteries, key, antenna and mast, and trekked back to the park. Fifteen minutes later, the antenna was standing proud and tall, and I made my first CQ. A security guard stopped by, and fearing the worst, I explained what I was doing. "Okay," she said, and drove away. A few minutes later I had a nice surprise—Rod Cerkoney, N0RC, showed up to operate with me!

The Tuna Tin 2 came back home, and I got it ready for the QRP Extravaganza Weekend (my name for it) on Halloween, with the QRP-ARCI/ARRL "Black Cat" party and the NorCal Zombie Shuffle operating event. You can read that tale in Rich Arland's "QRP Power" column in this issue.

Are We Having Fun Yet?

Did I have fun? Do you need to ask? I guess I was just in the right place at the right time, and have been privileged to be the center of all this Tuna Tin 2 activity. What is important to me, though, is that the magic that DeMaw created in the ARRL Lab still lives. It has, in fact, it has taken on a life of its own.

The Tuna Tin 2 will be on the air on 40 meters a lot over the rest of the winter, spring and summer. You'll hear it from W1RFI, from W1AW, and possibly some other station locations. I do have one more "special event" in the works, but I am sworn to secrecy. The Tuna Tin 2 will play a part in it. I won't tell you what call it will use, but I will say that you will

know it when you hear it. And when you do, you will know that the magic is still alive.

I hope that lots of hams build some of the various Tuna Tin 2 replicas, and that they get a chance to work the original. I will do my best to keep it on the air. I am sure that Doug DeMaw would approve.— *W1RFI*

[1]See the Arizona ScQRPions site on the Web at: **http://www.extremezone.com/~ki7mn/sqrppage.htm.**

[2]The DK9SQ mast is available for $99 plus $5 shipping and handling from Kanga US, 3521 Spring Lake Dr, Findlay, OH 45840; tel 419-423-4604; **kanga@bright.net**; **http://www.bright.net/~kanga/kanga/.**

[3]Colorado QRP Club, PO Box 371883, Denver CO 80237-1883; **rschneid@ix.netcom.com**; **http://www.cqc.org/.**

Ed Hare, W1RFI, operating the TT2 from his sister's apartment in Los Angeles.

TT2 Performance

Keying quality with this rig was good with several kinds of crystals tried. There was no sign of chirp. Without shaping, the keying is fairly hard (good for weak-signal work), but there were no objectionable clicks heard in the station receiver. There is a temptation among some QRP experimenters to settle for a one-transistor oscillator type of rig. For academic purposes, that kind of circuit is great. But, for on-the-air use, it's better to have at least two transistors. This isolates the oscillator from the antenna, thereby reducing harmonic radiation. Furthermore, the efficiency of oscillators is considerably lower than that of an amplifier. Many of the "yoopy" QRP CW signals on our bands are products of one-transistor crystal oscillators. Signal quality should be good, regardless of the power level used.

The voltages shown in Figure 1 will be helpful in troubleshooting this rig. All dc measurements were made with a VTVM. The RF voltages were measured with an RF probe and a VTVM, The values may vary somewhat, depending on the exact characteristics of the transistors chosen. The points marked 1 and 2 (in circles) can be opened to permit insertion of a dc milliammeter. This will be useful in determining the dc input power level for each stage. Power output can be checked by means of an RF probe from J2 to ground. Measurements should be made with a 51- or 56-Ω resistor as a dummy load. For 350 mW of output, there should be 4.4 V_{rms} across the 56-Ω resistor.

Operating voltage for the transmitter can be obtained from nine Penlite cells connected in series (13.5 volts). For greater power reserve one can use size C or D cells wired in series. A small ac-operated 12- or 13-V regulated dc supply is suitable also, especially for home-station work.—*W1FB*

[Although this rig met all the Part 97 surious emission requirements when built in 1976, additional filtering is needed to meet today's rules. A bandpass filter for 40 meters is shown as an inset in Figure 1. It can be installed between S1 and the antenna jack.—*W1RFI*]

Circuit Details

A look at Figure 1 will indicate that there's nobody at home, so to speak, in the two-stage circuit. A Pierce type of crystal oscillator is used at Q1. Its output tickles the base of Q2 (lightly) with a few mW of drive power, causing Q2 to develop approximately 450 mW of dc input power as it is driven into the Class C mode. Power output was measured as 350 mW (1/3 W), indicating an amplifier efficiency of 70%.

The collector circuit of Q1 is not tuned to resonance at 40 meters. L1 acts as an RF choke, and the 100-pF capacitor from the collector to ground is for feedback purposes only. Resonance is actually just below the 80-meter band. The choke value is not critical and could be as high in inductance as 1 mH, although the lower values will aid stability.

The collector impedance of Q2 is approximately 250 Ω at the power level specified. Therefore, T1 is used to step the value down to around 60 Ω (4:1 transformation) so that the pi network will contain practical values of L and C. The pi network is designed for low Q (loaded Q of 1) to assure ample bandwidth on 40 meters. This will eliminate the need for tuning controls. Since a pi network is a low-pass filter, harmonic energy is low at the transmitter output. The pi network is designed to transform 60 to 50 Ω.

L1 is a 22-μH molded inductor. L2 is made with 19 turns of #26 wire on a T-37-2 core. Final adjustment of this coil (L2) is done with the transmitter operating into a 50-Ω load. The coil turns are moved closer together or farther apart until maximum output is noted. The wire is then cemented in place by means of hobby glue or Q dope

T1 is made with 16 turns of #26 wire on the primary, 8 turns of #26 wire on the secondary, on an FT-37-43 ferrite core. This is good material for making broadband transformers, as very few wire turns are required for a specified amount of inductance, and the Q of the winding will be low (desirable).

Increased power can be had by making the emitter resistor of Q2 smaller in value. However, the collector current will rise if the resistor is decreased in value, and the transistor just might "go out for lunch," permanently, if too much collector current is allowed to flow. The current can be increased to 50 mA without need to worry, and this will elevate the power output to roughly 400 mW.

Construction Notes

The PC board can be cut to circular form by means of a nibbling tool or coping saw. It should be made so it just clears the inner diameter of the lip that crowns the container. The can is prepared by cutting the closed end so that 1/8 inch of metal remains all the way around the rim. This will provide a shelf for the circuit board to rest on. After checkout is completed, the board can be soldered to the shelf at four points to hold it in place. The opposite end of the can is open.

Summary Comments

Skeptics may chortle with scorn and amusement at the pioneer outlook of QRP enthusiasts. Their lack of familiarity with low-power operating may be the basis for their disdain. Those who have worked at micropower levels know that Worked All States is possible on 40 meters with less than a watt of RF energy. From the writer's location in Connecticut, all call areas of the USA have been worked at the 1/4-W power plateau. It was done with only a 40-meter coax-fed dipole, sloping to ground at approximately 45° from a steel tower. Signal reports ranged from RST 449 to RST 589, depending on conditions. Of course, there were many RST 599 reports too, but they were the exception rather than the rule. The first QSO with this rig came when Al, K4DAS, of Miami answered the writer's "CQ" at 2320 UTC on 7014 kHz. An RST 569 was received, and a 20-minute ragchew ensued. The copy at K4DAS was "solid."

If you've never tried QRP before, the first step is easy. Just contact the QRP Amateur Radio Club International (QRP-ARCI), 848 Valbrook Court, Lilburn, GA 30047-4280; **http://www.qrparci.org/**.

---

left-over parts from the assortments. Depending on how shrewd he is at the bargaining game, a flea-market denizen can probably put this unit together for a few bucks.

Fishy Excitement at the Meriden ARC

Renewed interest in the Tuna Tin 2 transceiver prompted the Meriden (Connecticut) Amateur Radio Club to build these classics as a club project. Bob Stephens, KB1CIW and Jamie Toole, N1RU secured components for 20 kits. Tim Mik, WY1U, supplied 20 cat food cans, cleaned and stripped of labels. (We had to assume that each can had, in fact, contained tuna flavor cat food. We didn't want to stray too far from the original design!) Tim also brought along his original Tuna Tin 2, which he had built as a newly licensed teenager over 20 years ago.

Several of the more experienced members were quite helpful in assisting those less knowledgeable in the arcane arts of schematic reading and toroid winding. Counting the number of turns, especially on the transformer, is not quite the simple task that it seems at first. Other tips on soldering and building in general were freely passed on from the veterans.

Honors for the first contact went to MARC president Bill Wawrzeniak, W1KKF. After finishing his rig, he brought it home, connected an antenna and almost immediately made contact with a California ham. With his new Tuna Tin 2, WY1U worked Ed Hare, W1RFI, operating the W1AW special event at ARRL HQ on Halloween. Most of the other kits were completed and put on the air over the next several weeks.

Building the Tuna Tin 2 is a terrific activity for any club. It can be completed in one or two evenings. The circuit is simple enough to provide an excellent springboard for education in electronic and RF theory without getting bogged down in too many esoteric topics. Building the kit is a great way to learn or sharpen construction skills. And, of course, there's no substitute for the pride and satisfaction of telling the station at the other end of the QSO, "RIG HR IS HMBRW TT2".—*John Bee, N1GNV, QST Advertising Manager*   CATS BY GIL, W1CJD

Receivers to Use With the QRP Transmitter

Perhaps you would like to build a more capable receiver than a crystal radio to go with the preceding transmitter. Or you might like a radio for general Amateur Radio reception. I will share several previously published receiver designs that have served their users well.

Simplest May be Best—an Amateur-Band Regenerative Receiver

The first receiver makes use of a *regenerative* detector. The regenerative detector is a kind of combination amplifying detector and beat oscillator, all in one. It can work well, but is a bit touchy to adjust. The following circuit—from *W1FB's QRP Notebook-2nd Edition*,[1] an ARRL publication by the late Doug DeMaw, probably represents the simplest receiver that could actually be used for amateur operation. See **Fig A1-19**.

Note that there are two tuning capacitors—C2 is the *bandset* capacitor, used to set the tuning range of the receiver to the frequency range of interest. C1 is the *bandspread* capacitor, used to provide fine tuning within a small segment of the tuning range. For successful operation both C1 and C2 should be solidly mounted so that they can't move. In addition, it will be best if each has a large knob with a mechanical reduction drive of some sort . The capacitors should mounted on a grounded metal panel so that the capacitance of the operator's hand doesn't change the frequency—these simple receivers can be touchy.

To receive radiotelegraph signals, the regeneration control, R1, is advanced until the detector just starts oscillating—you'll hear a difference in the sound at that point. At the point of oscillation, you will hear a nice beat note as you tune through a telegraph signal and the detector will be at its most sensitive and selective. It is also possible to hear AM signals (in the 41-meter shortwave broadcast band) by reducing the regeneration to the point at which the oscillation just stops. To operate with the Tuna

Tin 2, set C1 to the middle of its range and tune C2 until you find the transmitter signal. Then use C2 to tune on either side of your transmit signal until you find someone to communicate with.

Even Better—a Direct-Conversion Amateur-Band Receiver

A direct-conversion (D-C) receiver is nothing more than a crystal set with its own beat oscillator tuned to the receive frequency. This receiver design, from the 2005 *ARRL Handbook for Radio Communication*[2], makes use of a crystal-controlled beat oscillator and a commercial balanced mixer. The mixer is followed by a high-gain audio system that includes band-limiting active filtering. By using a crystal oscillator, made somewhat variable by C7 in the diagram, and selecting the same frequency as your transmitter's crystal, you are sure to have the receiver tuner to the proper frequency and to have a stable arrangement.

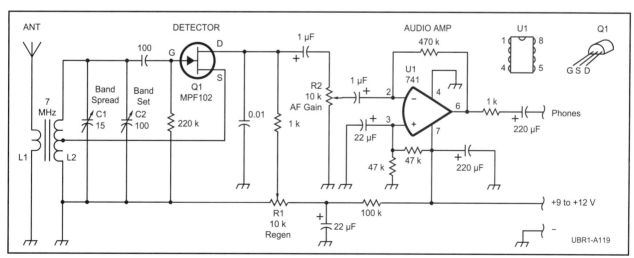

Fig A1-19—Schematic diagram of W1FB's simple regenerative 40-meter receiver that can be used with the "Tuna Tin 2" transmitter in Project 6. Polarized capacitors are 16-V or greater tantalum or electrolytic. Resistors are ¼-W carbon composition, except for R1 and R2, which are carbon composition controls (panel mounted). L2 is a 6-μH toroidal inductor. Use 35 turns of #28 enameled wire on an Amidon T50-2 toroid. Tap L2 at 8 turns above the grounded end. L1 has 3 turns of #28 enameled wire. Other VHF JFETs may be used at Q1, such as the 2N4416.

A Rock-Bending Receiver for 7 MHz

This simple receiver by Randy Henderson, WI5W, originally published in Aug 1995 *QST,* is a direct-conversion type that converts RF directly to audio. Building a stable oscillator is often the most challenging part of a simple receiver. This one uses a tunable crystal-controlled oscillator that is both stable and easy to reproduce. All of its parts are readily available from multiple sources and the fixed-value capacitors and resistors are common components available from many electronics parts suppliers.

The Circuit

This receiver works by mixing two radio-frequency signals together. One of them is the signal you want to hear, and the other is generated by an oscillator circuit (Q1 and associated components) in the receiver. In **Fig 14.61**, mixer U1 puts out sums and differences of these signals and their harmonics. We don't use the sum of the original frequencies, which comes out of the mixer in the vicinity of 14 MHz. Instead, we use the frequency *difference* between the incoming signal and the receiver's oscillator—a signal in the audio range if the incoming signal and oscillator frequencies are close enough to each other. This signal is filtered in U2, and amplified in U2 and U3. An audio transducer (a speaker or headphones) converts U3's electrical output to audio.

How the Rock Bender Bends Rocks

The oscillator is a tunable crystal oscillator—a variable crystal oscillator, or *VXO*. Moving the oscillation frequency of a crystal like this is often called *pulling*. Because crystals consist of precisely sized pieces of quartz, crystals have long been called *rocks* in ham slang—and receivers, transmitters and transceivers that can't be tuned around due to

crystal frequency control have been said to be *rockbound*. Widening this rockbound receiver's tuning range with crystal pulling made *rock bending* seem just as appropriate!

L2's value determines the degree of pulling available. Using FT-243-style crystals and larger L2 values, the oscillator reliably tunes from the frequency marked on the holder to about 50 kHz below that point with larger L2 values. (In the author's receiver a 25-kHz tuning range was achieved.) The oscillator's frequency stability is very good.

Inductor L2 and the crystal, Y1, have more effect on the oscillator than any other components. Breaking up L2 into two or three series-connected components often works better than using one RF choke. (The author used three molded RF chokes in series—two 10-µH chokes and one 2.7-µH unit.) Making L2's value too large makes the oscillator stop.

The author tested several crystals at Y1. Those in FT-243 and HC-6-style holders seemed more than happy to react to adjustment of C7 (TUNING). Crystals in the smaller HC-18 metal holders need more inductance at L2 to obtain the same tuning range. One tiny HC-45 unit from International Crystals needed 59 µH to eke out a mere 15 kHz of tuning range.

Input Filter and Mixer

C1, L1, and C2 form the receiver's input filter. They act as a peaked *low-pass* network to keep the mixer, U1, from responding to signals higher in frequency than the 40-meter band. (This is a good idea because it keeps us from hearing video buzz from local television transmitters, and signals that might mix with harmonics of the receiver's VXO.) U1, a Mini-Circuits SBL-1, is a passive diode-ring mixer. Diode-ring mixers usually perform better if the output is terminated properly. R11 and C8

provide a resistive termination at RF without disturbing U2A's gain or noise figure.

Audio Amplifier and Filter

U2A amplifies the audio signal from U1. U2B serves as an active low-pass filter. The values of C12, C13 and C14 are appropriate for listening to CW signals. If you want SSB stations to sound better, make the changes shown in the caption for Fig 14.61.

U3, an LM386 audio power amplifier IC, serves as the receiver's audio output stage. The audio signal at U3's output is more than a billion times more powerful than a weak signal at the receiver's input, so don't run the speaker/earphone leads near the circuit board. Doing so may cause a squealy audio oscillation at high volume settings.

Construction

If you're already an accomplished builder, you know that this project can be built using a number of construction techniques, so have at it! If you're new to building, you should consider building the Rock-Bending Receiver on a printed circuit (PC) board. (The parts list tells where you can buy one ready-made.) See **Fig 14.62** for details on the physical layout of several important components used in the receiver. **Fig 14.63** shows photos of two different receivers using two different approaches to construction—one using a PC board and the other using "ugly" techniques.

If you use a homemade double-sided circuit board based on the PC pattern on the accompanying CD, you'll notice that it has more holes than it needs to. The extra holes (indicated in the part-placement diagram with square pads) allow you to connect its ground plane to the ground traces on its foil side. (Doing

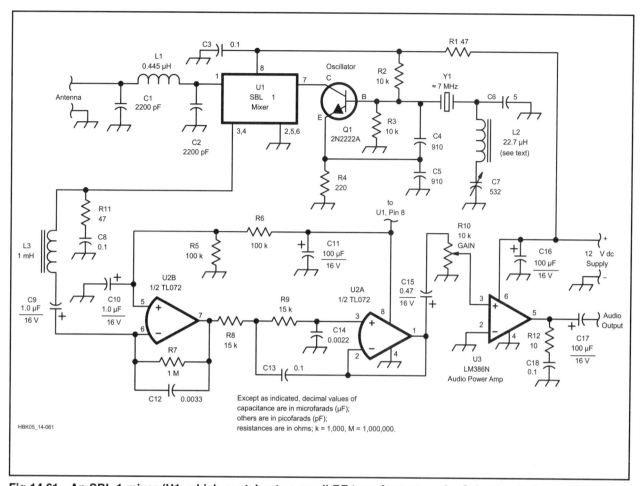

Fig 14.61—An SBL-1 mixer (U1, which contains two small RF transformers and a Schottky-diode quad), a TL072 dual op-amp IC (U2) and an LM386 low-voltage audio power amplifier IC (U3) do much of the Rock-Bending Receiver's magic. Q1, a variable crystal oscillator (VXO), generates a low-power radio signal that shifts incoming signals down to the audio range for amplification in U2 and U3. All of the circuit's resistors are ¼-W, 5%-tolerance types; the circuit's polarized capacitors are 16-V electrolytics, except C10, which can be rated as low as 10 V. The 0.1-µF capacitors are monolithic or disc ceramics rated at 16 V or higher.

C1, C2—Ceramic or mica, 10% tolerance.

C4, C5, and C6—Polystyrene, dipped silver mica, or C0G (formerly NP0) ceramic, 10% tolerance.

C7—Dual-gang polyethylene-film variable (266 pF per section) available as #24TR218 from Mouser Electronics (800-346-6873, 817-483-4422). Screws for mounting C7 are Mouser #48SS003. A rubber equipment foot serves as a knob. (Any variable capacitor with a maximum capacitance of 350 to 600 pF can be substituted; the wider the capacitance range, the better.)

C12, C13, C14—10% tolerance. For SSB, change C12, C13 and C14 to 0.001 µF.

U2—TL072CN or TL082CN dual JFET op amp.

L1—4 turns of AWG #18 wire on ³/₄-inch PVC pipe form. Actual pipe OD is 0.85 inch. The coil's length is about 0.65 inch; adjust turns spacing for maximum signal strength. Tack the turns in place with cyanoacrylic adhesive, coil dope or Duco cement. (As a substitute, wind 8 turns of #18 wire around 75% of the circumference of a T-50-2 powdered-iron core. Once you've soldered the coil in place and have the receiver working, expand and compress the coil's turns to peak incoming signals, and then cement the winding in place.)

L2—Approximately 22.7 µH; consists of one or more encapsulated RF chokes in series (two 10-µH chokes

[Mouser #43HH105 suitable] and one 2.7-µH choke [Mouser #43HH276 suitable] used by author). See text

L3—1-mH RF choke. As a substitute, wind 34 turns of #30 enameled wire around an FT-37-72 ferrite core.

Q1—2N2222, PN2222 or similar small-signal, silicon NPN transistor.

R10—5 or 10-kΩ audio-taper control (RadioShack No. 271-215 or 271-1721 suitable).

U1—Mini-Circuits SBL-1 mixer.

Y1—7-MHz fundamental-mode quartz crystal. Ocean State Electronics carries 7030, 7035, 7040, 7045, 7110 and 7125-kHz units.

PC boards for this project are available from FAR Circuits.

so reduces the inductance of some of the board's ground paths.) Pass a short length of bare wire (a clipped-off component lead is fine) into each of these holes and solder on both sides. Some of the circuit's components (C1, C2 and others) have grounded leads accessible on both sides of the board. Solder these leads on both sides of the board.

Another important thing to do if you use a homemade double-sided PC board is to countersink the ground plane to clear all ungrounded holes. (Countersinking clears copper away from the holes so components won't short-circuit to the ground plane.) A $1/4$-inch-diameter drill bit works well for this. Attach a control knob to the bit's shank and you can safely use the bit as a manual countersinking tool. If you countersink your board in a drill press, set it to about 300 rpm or less, and use very light pressure on the feed handle.

Mounting the receiver in a metal box or cabinet is a good idea. Plastic enclosures can't shield the TUNING capacitor from the presence of your hand, which may slightly affect the receiver tuning. You don't have to completely enclose the receiver—a flat aluminum panel screwed to a wooden base is an acceptable alternative. The panel supports the tuning capacitor, GAIN control and your choice of audio connector. The

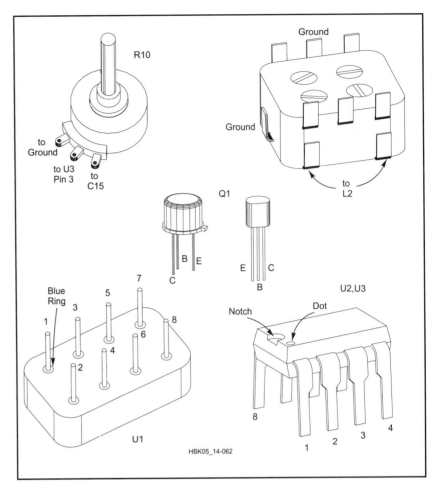

Fig 14.62—The Mouser Electronics part suggested for C7 has terminal connections as shown here. (You can use any variable capacitor with a maximum capacitance of 350 to 600 pF for C7, but its terminal configuration may differ from that shown here.) Two Q1-case styles are shown because plastic or metal transistors will work equally well for Q1. If you build your Rock-Bending Receiver using a prefab PC board, you should mount the ICs in 8-pin mini-DIP sockets rather than just soldering the ICs to the board.

Fig 14.63—Ground-plane construction, PC-board construction—both of these approaches can produce the same good Rock Bending Receiver performance. (WI5W built the one that looks nice, and ex-W9VES—who wrote this caption—built the one that doesn't.)

base can support the circuit board and antenna connector.

Checkout

Before connecting the receiver to a power source, thoroughly inspect your work to spot obvious problems like solder bridges, incorrectly inserted components or incorrectly wired connections. Using the schematic (and PC-board layout if you built your receiver on a PC board), recheck every component and connection one at a time. If you have a digital voltmeter (DVM), use it to measure the resistance between ground and everything that should be grounded. This includes things like pin 4 of U2 and U3, pins 2, 5, 6 of U1, and the rotor of C7.

If the grounded connections seem all right, check some supply-side connections with the meter. The connection between pin 6 of U3 and the positive power-supply lead should show less than 1 W of resistance. The resistance between the supply lead and pin 8 of U1 should be about 47 W because of R1.

If everything seems okay, you can apply power to the receiver. The receiver will work with supply voltages as low as 6 V and as high as 13.5 V, but it's best to stay within the 9 to 12-V range. When first testing your receiver, use a current-limited power supply (set its limiting between 150 and 200 mA) or put a 150-mA fuse in the connection between the receiver and its power source. Once you're sure that everything is working as it should, you can remove the fuse or turn off the current limiting.

If you don't hear any signals with the antenna connected, you may have to do some troubleshooting. Don't worry; you can do it with very little equipment.

Trouble?

The first clue to look for is noise. With the GAIN control set to maximum, you should hear a faint rushing sound in the speaker or headphones. If not, you can use a small metallic tool and your body as a sort of test-signal generator. (If you have any doubt about the safety of your power supply, power the Rock-Bending Receiver from a battery during this test.) Turn the GAIN control to maximum. Grasp the metallic part of a screwdriver, needle or whatever in your fingers, and use the tool to touch pin 3 of U3. If you hear a loud scratchy popping sound, that stage is working. If not, then something directly related to U3 is the problem.

You can use this technique at U2 (pin 3, then pin 5) and all the way to the antenna. If you hear loud pops when touching either end of L3 but not the antenna connector, the oscillator is probably not working. You can check for oscillator activity by putting the receiver near a friend's transceiver (both must be in the same room) and listening for the VXO. Be sure to adjust the TUNING control through its range when checking the oscillator.

The dc voltage at Q1's base (measured without the RF probe) should be about half the supply voltage. If Q1's collector voltage is about equal to the supply voltage, and Q1's base voltage is about half that value, Q1 is probably okay. Reducing the value of L2 may be necessary to make some crystals oscillate.

Operation

Although the Rock-Bending Receiver uses only a handful of parts and its features are limited, it performs surprisingly well. Based on tests done with a Hewlett-Packard HP 606A signal generator, the receiver's minimum discernible signal (by ear) appears to be 0.3 µV. The author could easily copy 1-µV signals with his version of the Rock-Bending Receiver.

Although most HF-active hams use transceivers, there are advantages in using separate receivers and transmitters. This is especially true if you are trying to assemble a simple home-built station.

Glossary

A-scope—A radar indicator showing range and target return amplitude.

Access code—one or more numbers and/or symbols that are keyed into the repeater with a DTMF tone pad to activate a repeater function, such as an autopatch.

Access Point (AP)—A wireless bridging device that connects 802.11 stations to shared resources and a wired network such as the Internet.

ACK—Acknowledgment, the control signal sent to indicate the correct receipt of a transmission block.

Active Region—The region in the characteristic curve of an analog device in which the signal is amplified linearly.

Address—A character or group of characters that identifies a source or destination.

Ad Hoc—In wireless LAN (WLAN) networks this is a direct wireless connection between two laptop computers without the use of an AP.

Admittance (Y)—The reciprocal of impedance, measured in siemens (S).

ADV (Amateur digital video)—A mode of operation in which *Amateur Radio* operators exchange video motion images using their personal computers.

AGC—Automatic gain control. AGC adjusts the gain of a receiver to compensate for received signal strength.

AFSK—Audio frequency-shift keying.

Alternating current—A flow of charged particles through a conductor, first in one direction, then in the other direction.

ALE—Automatic link establishment. A system that automatically sets the frequency of HF radios to the optimum of available channels.

AM (Amplitude modulation)—The oldest voice operating mode still found on the amateur bands. The more common voice mode, *SSB*, is actually a narrower-bandwidth variation of AM.

Amateur Radio—A radiocommunication service for the purpose of self training, intercommunication and technical investigations carried out by amateurs, that is, duly authorized persons interested in radio technique solely with a personal aim and without pecuniary interest. (*Pecuniary* means payment of any type, whether money or goods.) Also called *ham radio*.

Amateur Radio operator—A person holding a license to operate a ham radio station.

Amateur Radio station—A station licensed in the amateur service, including necessary equipment.

Amateur Service—A radio communication service for the purpose of self training, intercommunication and technical investigations carried out by amateurs, that is, duly authorized persons interested in radio technique solely with a personal aim and without pecuniary interest.

Ampere—A measure of flow of charged particles per unit time. One ampere represents one coulomb of charge flowing past a point in one second.

AMPS (Advanced Mobile Phone Service)—First standardized cellular service in the world, released in 1983. Uses the 800-900 MHz frequency band.

AMRAD—Amateur Radio Research and Development Corporation, a nonprofit organization involved in packet-radio development.

AMSAT (Radio Amateur Satellite Corporation)—An international membership organization that designs, builds and promotes the use of Amateur Radio satellites.

AMTOR—Amateur teleprinting over radio, an amateur radioteletype transmission technique employing error correction as specified in several CCIR documents 476-2 through 476-4 and 625. CCIR Rec. 476-3 is reprinted in the *Proceedings* of the Third ARRL Amateur Radio Computer Networking Conference, available from ARRL Hq.

Analog—A signal that can vary continuously between a maximum and minimum value. For example, the voice voltage waveform from the output of a microphone is analog. RF voltage waveforms (as those from AM, FM and SSB transmitters) are also analog.

Angle Modulation—A term referring to frequency or phase modulation.

Anode—The element of an analog device that accepts electrons.

ANSI—American National Standards Institute

Answer—The station intended to receive a call. In modem usage, the called station or modem tones associated therewith.

AP—Access point.

ARES®—An ARRL program specializing in emergency communication.

ARISS—An acronym for Amateur Radio on the International Space Station.

ARQ—Automatic repeat request, an error-sending station, after transmitting a data block, awaits a reply (ACK or NAK) to determine whether to repeat the last block or proceed to the next.

Array—An antenna consisting of multiple elements designed to operate together to result in a particular direction of radiation.

ARRL—The membership organization for Amateur Radio operators in the US.

ASCII—American National Standard Code for Information Interchange, a code consisting of seven information bits.

Atom—The smallest particle of matter that makes up an element. Consists of protons and neutrons in the central area called the nucleus, with electrons surrounding this central region.

ATV (Amateur television)—A mode of operation that amateur radio operators use to exchange pictures from their ham stations.

Aurora—Related to the northern lights or aurora borealis phenomenon. The aurora results in localized ionization that allows over the horizon radio propagation, particularly in the VHF region.

Autopatch—a device that interfaces a repeater to the telephone system to permit repeater users to make telephone calls. Often just called a "patch."

Auxiliary station—An amateur station, other than in a message-forwarding system, transmitting communication point-to-point within a system of cooperating amateur stations.

AX.25—Amateur packet-radio link-layer protocol. Copies of protocol specification are available from ARRL HQ.

B-scope—A radar indicator showing azimuth, range and target return amplitude. See PPI.

Back porch—The blank part of a scan line immediately following the horizontal sync pulse.

Backwave—An unwanted signal emitted between the pulses of an on/off-keyed signal.

Balanced—A relationship in which two stations communicate with one another as equals; that is, neither is a primary (master) or secondary (slave).

Band—A range of frequencies. Hams are authorized to transmit on many different bands.

Bandwidth—The width of a frequency band outside of which the mean power of the transmitted signal is attenuated at least 26 dB below the mean power of the transmitted signal within the band.

Base—The middle layer of a bipolar transistor, often the input.

Baseband—An information signal, often made up of multiple combined signals (see "multiplex"), that is used to modulate a radio system for transmission.

Baud—A unit of signaling speed equal to the number of discrete conditions or events per second. (If the duration of a pulse is 20 ms, the signaling rate is 50 bauds or the reciprocal of 0.02, abbreviated Bd).

Baudot code—A coded character set in which five bits represent one character. Used in the US to refer to ITA2.

Beacon—An amateur station transmitting communication for the purposes of observation of propagation and reception or other related experimental activities.

Beam antenna—A type of radio antenna that can be pointed in any direction.

Beat frequency oscillator (BFO)—An oscillator in a receiver designed to allow detection of suppressed carrier or radiotelegraph signals.

BER—Bit error rate.

BERT—Bit-error-rate test.

Biasing—The addition of a dc voltage or current to a signal at the input of an analog device, which changes the signal's position on the characteristic curve.

Bipolar Transistor—An analog device made by sandwiching a layer of doped semiconductor between two layers of the opposite type: PNP or NPN.

Bistatic—A radar system in which transmitting and receiving locations are in different locations.

Bit stuffing—Insertion and deletion of 0s in a frame to preclude accidental occurrences of flags other than at the beginning and end of frames.

Bit—Binary digit, a single symbol, in binary terms either a one or zero.

Bleeder—A resistive load across the output or filter of a power supply, intended to quickly discharge stored energy once the supply is turned off.

Break—the word used to interrupt a conversation on a repeater *only* to indicate that there is an emergency.

Break-in—The ability to hear between elements or words of a keyed signal.

Broadcasting—Transmissions intended for reception by the general public, either direct or relayed.

Buffer—An analog stage that prevents loading of one analog stage by another.

Bypass capacitor—a capacitor used to provide a low-impedance radio-frequency path around a circuit element.

Byte—A group of bits, usually eight.

Call sign—A series of unique letters and numbers assigned to a person who has earned an Amateur Radio license.

Capacitance (C)—The ability to store electrical energy in an electrostatic field, measured in farads (F). A device with capacitance is a capacitor.

Carrier-operated relay (COR)—a device that causes a repeater to transmit in response to a received signal.

Carrier Sense Multiple Access / Collision Avoidance (CSMA/CA)—The wireless method that tries to avoid simultaneous access or collisions by not transmitting, if another signal is detected on the same frequency channel.

Carrier power—The average power supplied to the antenna transmission line by a transmitter during one RF cycle taken under the condition of no modulation.

Cascade—Placing one analog stage after another to combine their effects on the signal.

Cathode—The element of an analog device that emits electrons.

CCIR—International Radio Consultative Committee, an International Telecommunication Union (ITU) agency.

CCITT—International Telegraph and Telephone Consultative Committee, an ITU agency. CCIR and CCITT recommendations are available from the UN Bookstore.

CDMA (Code Division Multiple Access)—A digital radio system that separates users by digital codes.

Cellular—Characteristic of or pertaining to a system of wireless communication made up of many individual cell units. The term itself is derived from the typical geographic honeycomb shape of the areas into which a coverage region is divided.

Certification—An equipment authorization granted by the FCC. It is used to ensure that equipment will function properly in the service for which it has been accepted. Most amateur equipment does not require FCC Certification, although HF power amplifiers and amplifier kits do. Part 15 Rules require FCC Certification for all receivers operating anywhere between 30 and 960 MHz. Amateur transmitters may not be legally used in any other service that requires FCC equipment authorization. For example, it is illegal to use a modified amateur transmitter in the police, fire or business radio services.

Channel—the pair of frequencies (input and output) used by a repeater.

Characteristic Curve—A plot of the relative responses of two or three analog-device parameters, usually output with respect to input.

Chirp—Incidental frequency modulation of a carrier as a result of oscillator instability during keying.

Chrominance—The color component of a video signal. NTSC and PAL transmit color images as a black-and-white compatible luminance signal along with a color subcarrier. The subcarrier phase represents the hue and the subcarrier's amplitude is the saturation. Robot color modes transmit pixel values as luminance (Y) and chrominance (R-Y [red minus luminance] and B-Y [blue minus luminance]) rather than RGB (red, green, blue).

Circular Mils—A convenient way of expressing the cross-sectional area of a round conductor. The area of the conductor in circular mils is found by squaring its diameter in mils (thousandths of an inch), rather than squaring its radius and multiplying by pi. For example, the diameter of 10-gauge wire is 101.9 mils (0.1019 inch). Its cross-sectional area is 10380 CM, or 0.008155 square inches.

Clamping—A nonlinearity in amplification where the signal can be made no larger.

Closed repeater—a repeater whose access is limited to a select group (see *open repeater*).

COFDM—Coded Orthogonal Frequency Division Multiplex, OFDM plus coding to provide error correction and noise immunity.

Collector—One of the outer layers of a bipolar transistor, often the output.

Collision—A condition that occurs when two or more transmissions occur at the same time and cause interference to the intended receivers.

Common-mode signals—signals that are in phase on both (or several) conductors in a system.

Compensation—The process of counteracting the effects of signals that are inadvertently fed back from the output to the input of an analog system. The process increases stability and prevents oscillation.

Conductance (G)— The reciprocal of resistance, measured in siemens (S).

Conducted signals—signals that travel by electron flow in a wire or other conductor.

Constellation—A set of points in the complex plane which represent the various combinations of phase and amplitude in a QAM or other complex modulation scheme.

Contact—A two-way communication between Amateur Radio operators.

Contention—A condition on a communications channel that occurs when two or more stations try to transmit at the same time.

Contest—An Amateur Radio activity in which hams and their stations compete to contact the most stations within a designated time period.

Control operator—An amateur operator designated by the licensee of a station to be responsible for the transmissions from that station to assure compliance with the FCC Rules.

Control point—The location at which the control operator function is performed.

Core Saturation (Magnetic)—That condition whereby the magnetic flux in a transformer or inductor core is more than the core can handle. If the flux is forced beyond this point, the permeability of the core will decrease, and it will approach the permeability of air.

Coulomb—A unit of measure of a quantity of electrically charged particles. One coulomb is equal to $6.25 \times 10_{18}$ electrons.

Courage Handi-Ham System—Membership organization for ham radio enthusiasts with various physical disabilities and abilities.

Courtesy beep—an audible indication that a repeater user may go ahead and transmit.

Coverage—the geographic area within which the repeater provides communications.

C-Rate—The charging rate for a battery, expressed as a ratio of the battery's ampere-hour rating.

CRC—Cyclic redundancy check, a mathematical operation. The result of the CRC is sent with a transmission block. The receiving station uses the received CRC to check transmitted data integrity.

Crowbar—A last-ditch protection circuit included in many power supplies to protect the load equipment against failure of the regulator in the supply. The crowbar senses an overvoltage condition on the supply's output and fires a shorting device (usually an SCR) to directly short-circuit the supply's output and protect the load. This causes very high currents in the power supply, which blow the supply's input-line fuse.

CSMA—Carrier sense multiple access, a channel access arbitration scheme in which packet-radio stations listen on a channel for the presence of a carrier before transmitting a frame.

CSMA/CD (Carrier Sense Multiple Access / Collision Detection)—A set of rules that determine how network devices respond when two devices attempt to use a data channel simultaneously (called a *collision*). After detecting a collision, a device waits a random delay time and then attempts to re-transmit the message. If the device detects a collision again, it now waits twice as long to try to re-transmit the message.

CTCSS—abbreviation for continuous tone-controlled squelch system, a series of subaudible tones that some repeaters use to restrict access. (see **closed repeater**).

Current (I)—The rate of electron flow through a conductor, measured in amperes (A).

Cutoff Region—The region in the characteristic curve of an analog device in which there is no current through the device. Also called the OFF region.

CW—Abbreviation for *continuous wave;* another name for *Morse code* telegraphy by radio. Also, International Morse code telegraphy emissions having designators with A, C, H, J or R as the first symbol; 1 as the second symbol; A or B as the third symbol; and emissions J2A and J2B.

Darlington Transistor—A package of two transistors in one case, with the collectors tied together, and the emitter of one transistor connected to the base of the other. The effective current gain of the pair is approximately the product of the individual gains of the two devices.

DARPA—Defense Advanced Research Projects Agency; formerly ARPA, sponsors of ARPANET.

Data—Telemetry, telecommand and computer communication emissions having designators with A, C, D, F, G, H, J or R as the first symbol; 1 as the second symbol; D as the third symbol; and emission J2D. Only a digital code of a type specifically authorized in this Part may be transmitted.

Data set—Modem.

dBd—Decibels with a reference to a dipole antenna. A way of indicating antenna gain in comparison to a dipole antenna. Typically 2 dB less than dBi.

dBi—Decibels with a reference to an ideal isotropic antenna. A way of indicating antenna gain in comparison to an antenna with uniform radiation in all directions.

DC-DC Converter—A circuit for changing the voltage of a dc source to ac, transforming it to another level, and then rectifying the output to produce direct current.

DCE—Data circuit-terminating equipment. The equipment (for example, a modem) that provides communication between the DTE and the line radio equipment.

DDS—Direct digital synthesis. The generation of a sinusoidal signal via a computer simulation of the waveform.

Decibel (dB)—a logarithmic unit of relative power measurement that expresses the ratio of two power levels.

Demodulator—A device that extracts the information waveform or baseband data from an radio signal.

Deviation—The maximum change of a carrier frequency under frequency or phase modulation.

DHCP (Dynamic Host Configuration Protocol)—An external assignment mechanism that provides a "care-of address" to a mobile client (see also *Foreign Agent*).

Differential-mode signals—Signals that arrive on two or more conductors such that there is a 180° phase difference between the signals on some of the conductors.

Digipeater—a packet radio (digital) repeater.

Digital—A signal that has only discrete values, usually two (logic 1 and logic 0), that changes at predetermined intervals. The value (e.g., voltage) present in a single time period is called a bit. The number of bits transferred per second is called the bit rate that has units of bits per second (bit/s), or kilobits per second (kbit/s), etc.

Digital communication—Computer-based communication modes such as PSK31, *packet radio* and *HSMM*.

Diode—A two-element vacuum tube or semiconductor with only a cathode and an anode (or plate).

Dipole antenna—A wire antenna often used on the high-frequency amateur bands.

Direct current—A flow of charged particles through a conductor in one direction only.

Direct-Sequence Spread Spectrum (DSSS)—The type of modulation used in 802.11b that is capable of maximum half-duplex data speeds of 11 Mbps.

Discriminator—A circuit used to convert an FM signal to baseband or audio.

Drain—The connection at one end of a field-effect-transistor channel, often the output.

DRM—Digital Radio Mondiale. A consortium of broadcasters, manufacturers, research and governmental organizations who are developing a system for digital broadcasting in the AM bands between 100 kHz and 30 MHz. The term is also used to refer to the broadcasts themselves.

DSP (Digital signal processing)—A newer technology that allows software to replace electronic circuitry.

DTMF—abbreviation for dual-tone multifrequency, the series of tones generated from a keypad on a ham radio transceiver (or a regular telephone).

Duplex or **full duplex**—a mode of communication in which a user transmits on one frequency and receives on another frequency simultaneously (see *half duplex*).

Duplexer—a device that allows the repeater transmitter and receiver to use the same antenna simultaneously.

DX—A ham radio abbreviation for *distance* or *foreign countries.*

DXCC—A popular ARRL award earned for contacting Amateur Radio operators in 100 different countries.

DX PacketCluster—A method of informing hams, via their computers, about the activities of stations operating from unusual locations.

DXpedition—A trip to an unusual location, such as an uninhabited island or other geographical or political entity which has few, if any, Amateur Radio operators, where hams operate while visiting. DXpeditions provide sought after contacts for hams who are anxious to have a radio contact with someone in a rare location.

Dynamic range—The difference between the largest and the smallest signal a system can process. A measure of receiver performance—how well it responds to weak signals in the presence of strong signals on adjacent channels.

EIA—Electronic Industries Alliance.

EIA-232-C—An EIA standard physical level interface between DTE (terminal) and DCE (modem), using 25-pin connectors.

Electromagnetic compatibility (EMC)—the ability of electronic equipment to be operated in its intended electromagnetic environment without either causing interference to other equipment or systems, or suffering interference from other equipment or systems.

Electromagnetic interference (EMI)—any electrical disturbance that interferes with the normal operation of electronic equipment.

Electron—A subatomic particle that has a negative charge and is the basis of electrical current.

Elmer—A traditional term for someone who enjoys helping newcomers get started in ham radio. A mentor.

E-mail—Electronic mail sent and received via computers with modems. Transmission media can be existing telephone or other communication lines, wireless, or not uncommonly—both.

Emergency communication—Amateur Radio communication that take place during a situation where there is danger to lives or property.

EMF—Electromotive Force is the term used to define the force of attraction between two points of different charge potential. Also called voltage.

Emission—Electromagnetic energy propagated from a source by radiation.

Emitter—One of the outer layers of a bipolar transistor, often the reference.

Encode—The process whereby a transmission contains additional data or code added to facilitate proper routing of the transmission to the desired point or points.

Encryption—Technology used to form a secure channel between a wireless client and the server to support user authentication, data integrity, and data privacy.

Energy—Capability of doing work. It is usually measured in electrical terms as the number of watts of power consumed during a specific period of time, such as watt-seconds or kilowatt-hours.

Envelope-delay distortion—In a complex waveform, unequal propagation delay for different frequency components.

Equalization—Correction for amplitude-frequency and/or phase-frequency distortion.

ERP—Effective radiated power. The power radiated by an antenna in a particular direction compared to a reference omnidirectional antenna.

ESN (Electronic Serial Number)—A manufacturer-assigned identity contained in a data transmission from a call placed to verify that the hardware used belongs to a valid cellular account.

External RF power amplifier—A device capable of increasing power output when used in conjunction with, but not an integral part of, a transmitter.

External RF power amplifier kit—A number of electronic parts, which, when assembled, is an external RF power amplifier, even if additional parts are required to complete assembly.

Eye pattern—An oscilloscope display in the shape of one or more eyes for observing the shape of a serial digital stream and any impairments.

EZNEC—One of a number of antenna performance prediction programs.

Fast Recovery Rectifier—A specially doped rectifier diode designed to minimize the time necessary to halt conduction when the diode is switched from a forward-biased state to a reverse-biased state.

FCC (Federal Communications Commission)—The government agency that regulates Amateur Radio in the US.

FCS—Frame check sequence. (See CRC.)

FDM (Frequency Division Multiplex)—A system that combines multiple voice or data streams into a single baseband signal for transmission.

FDMA (Frequency Division Multiple Access)—A radio system that separates user channels by frequency. Amateur Radio equipment presently uses FDMA.

FEC—Forward error correction, an error control technique in which the transmitted data is sufficiently redundant to permit the receiving station to correct some errors.

Field—Collection of top to bottom scan lines. When interlaced, a field does not contain adjacent scan lines and there is more than one field per frame.

Field & Educational Services (F&ES)—Staff at ARRL Headquarters that helps newcomers get started in ham radio and supports hams who help newcomers.

Field Day—A popular Amateur Radio activity during which hams set up radio stations outdoors and away from electrical service to simulate emergencies.

Field-Effect Transistor (FET)—An analog device with a semiconductor channel whose width can be modified by an electric field. Also called a unipolar transistor.

Field Organization—A cadre of ARRL volunteers who perform various services for the Amateur Radio community at the local and state level.

Filter—a network of resistors, inductors and/or capacitors that offer little resistance to certain frequencies while blocking or attenuating other frequencies.

Flux density (B)—The number of magnetic-force lines per unit area, measured in gauss.

Foldback Current Limiting—A special type of current limiting used in linear power supplies, which reduces the current through the supply's regulator to a low value under short circuited load conditions in order to protect the series pass transistor from excessive power dissipation and possible destruction.

Footprint—The coverage area of an individual cell.

Foreign Agent—A special "node" that is present on a foreign network and provides mobility services to visiting mobile nodes.

FOT—Frequency of optimum transmission. This is a frequency, between the lowest useable frequency (LUF) and maximum useable frequency (MUF), that will likely provide the best signal to noise ratio for a particular HF radio circuit.

FM (Frequency modulation)—An operating *mode* commonly used on ham radio *repeaters*.

Fox hunt—A competitive Amateur Radio activity in which hams track down a transmitted signal.

Frequency (f)—The rate of change of an ac voltage or current, measured in cycles per second, or hertz (Hz).

Frequency coordinator—An entity, recognized in a local or regional area by amateur operators whose stations are eligible to be auxiliary or repeater stations, that recommends transmit/receive channels and associated operating and technical parameters for such stations in order to avoid or minimize potential interference.

Frequency Hopping Spread Spectrum (FHSP)—A type of modulation used in early 802.11 devices that uses a time-varied narrow signal to spread the signal over a wide band. Maximum half-duplex data rate is 2 Mbps.

Front porch—The blank part of a television scan line just before the horizontal sync.

FSK—Frequency-shift keying.

FSTV (Fast-scan television)—A mode of operation that Amateur Radio operators can use to exchange live TV images from their stations. Same as common, full-color, motion commercial broadcast TV.

Full quieting—a received signal that contains no noise.

Fundamental overload—interference resulting from the fundamental signal of a radio transmitter.

Gain—see *Amplification*.

Gain-Bandwidth Product—The interrelationship between amplification and frequency that defines the limits of the ability of a device to act as a linear amplifier. In many amplifiers, gain times bandwidth is approximately constant.

Gate—The connection at the control point of a field-effect transistor, often the input.

Grid—The vacuum-tube element that controls the electron flow from cathode to plate. Additional grids in some tubes perform other control functions to improve performance.

GPS (Global Positioning System)—A Department of Defense-developed, worldwide, satellite-based radio navigation system.

Ground—a low-impedance electrical connection to the earth. Also, a common reference point in electronic circuits.

Ground Fault (Circuit) Interrupter (GFI or GFCI)—A safety device installed between the household power mains and equipment where there is a danger of personnel touching an earth ground while operating the equipment. The GFI senses any current flowing directly to ground and immediately switches off all power to the equipment to minimize electrical shock. GFIs are now standard equipment in bathroom and outdoor receptacles.

Group—The multiplexed combination of 12 voice channels into a baseband signal of 48 kHz bandwidth.

GSM—A cellular mobile telephone system used in many non-US countries.

Half duplex—a mode of communication in which a user transmits at one time and receives at another time.

Ham band—A range of frequencies on which ham communication are authorized.

Ham radio—Another name for *Amateur Radio*.

Handoff—Process whereby a mobile telephone network automatically transfers a call from cell to cell—possibly to another channel—as a mobile crosses adjacent cells.

Harmful interference—Interference which endangers the functioning of a radionavigation service or of other safety services or seriously degrades, obstructs or repeatedly interrupts a radiocommunication service—including ham radio—operating in accordance with the international Radio Regulations.

Harmonics—signals at exact integral multiples of the operating (or *fundamental*) frequency.

Heterodyne—A system in which a locally generated signal is used to process a signal to result in a translation of the information content to a different frequency.

High-pass filter—a filter designed to pass all frequencies above a cutoff frequency, while rejecting frequencies below the cutoff frequency.

HF (High frequency)—The radio frequencies from 3 to 30 MHz.

Hole—A positively charged "particle" that results when an electron is removed from an atom in a semiconductor crystal structure.

Host—As used in packet radio, a computer with applications programs accessible by remote stations.

HSMM (High Speed Multimedia)—A digital radio communication technique using spread spectrum modes primarily on UHF to simultaneously send and receive video, voice, text, and data.

IA5—International Alphabet No. 5, a 7-bit coded character set, CCITT version of ASCII.

IARU (International Amateur Radio Union)—The international organization made up of national Amateur Radio organizations such as the ARRL.

IBOC—In Band On Channel. A method of using the same channel on the AM or FM broadcast bands to transmit simultaneous digital and analog modulation.

IF—Intermediate frequency. A frequency used to process signals within a receiver or transmitter following heterodyning from the carrier frequency.

IFF—Identification friend or foe. A system developed during WWII to allow radio identification of distant aircraft to determine military aircraft identity. This system has been expanded to provide identification and altitude data about civilian aircraft to air traffic control systems.

Image—A signal observed in a receiver as a result of undesired products from a heterodyne mixer.

Immunity—the ability of electronic equipment to reject interference from external sources of electromagnetic energy. This is the conjugate of the term "susceptibility" and is the term typically used in the commercial world.

Impedance (Z)—The complex combination of resistance and reactance, measured in ohms (Ω).

Induction—the transfer of electrical signals via magnetic coupling.

Inductance (L)—The ability to store electrical energy in a magnetic field, measured in henrys (H). A device, such as a coil with inductance, is an inductor.

Information bulletin—A message directed only to amateur operators consisting solely of subject matter of direct interest to the amateur service.

Information field—Any sequence of bits containing the intelligence to be conveyed.

Input-Output Differential—The voltage drop appearing across the series pass transistor in a linear voltage regulator. This term is usually stated as a minimum value, which is that voltage necessary to allow the regulator to function and conduct current. A typical figure for this drop in most three-terminal regulator ICs is about 2.5 V. In other words, a regulator that is to provide 12.5 V dc will need a source voltage of at least 15.0 V at all times to maintain regulation.

Institute of Electrical and Electronics Engineers (IEEE)—The professional standards setting organization for data networking devices.

Integrated Circuit (IC)—A semiconductor device in which many components, such as diodes, bipolar transistors, field-effect transistors, resistors and capacitors are fabricated to make an entire circuit.

Interference—the unwanted interaction between electronic systems.

Interlace—Scan line ordering other than the usual sequential top to bottom. For example, NTSC sends a field with just the even lines in 1/60 second, then a field with just the odd lines in 1/60 second. This results in a complete frame 30 times a second. AVT "QRM" mode is the only SSTV mode that uses interlacing.

Intermodulation—the undesired mixing of two or more frequencies in a nonlinear device, which produces additional frequencies.

Ionosphere—A region extending above the earth's surface from a distance of about 30 to 250 miles. This region is ionized by solar radiation can refract radio waves depending on frequency and ionospheric conditions.

Inverter—A circuit for producing ac power from a dc source.

Iridium—A commercial low earth orbiting communications satellite system.

ISB—A system in which the channel bandwidth allocated to a double sideband AM signal is used to carry two independent signals, one on each sideband.

ISI—Intersymbol interference; slurring of one symbol into the next as a result of multipath propagation.

ISO—International Organization for Standardization.

International Morse code—A dot-dash code as defined in International Telegraph and Telephone Consultative Committee (CCITT) Recommendation F.1 (1984), Division B, I. Morse Code.

Isotropic—An antenna that radiates equally in all directions. An idealized antenna used as a reference for real antenna that don't.

ITA2—International Telegraph Alphabet No. 2, a CCITT 5-bit coded character set commonly called the Baudot or Murray code.

ITU (International Telecommunication Union)—An agency of the United Nations that allocates the radio spectrum among the various radio services.

Jitter—Unwanted variations in amplitude or phase in a digital signal.

Joule—Measure of a quantity of energy. One joule is defined as one Newton (a measure of force) acting over a distance of one meter.

Junction FET (JFET)—A field-effect transistor that forms its electric field across a PN junction.

Key clicks—Undesired switching transients beyond the necessary bandwidth of a Morse code transmission caused by improperly shaped modulation envelopes.

LAP—Link access procedure, CCITT X.25 unbalanced-mode communications.

LAPB—Link access procedure, balanced, CCITT X.25 balanced-mode communications.

LEO—Low earth orbit. A term referring to satellites that are in orbits well below the 22,400 mile geostationary orbital position.

Line Sequential—A method of color SSTV transmission that sends red, green, and blue information for each sequential scan line. This approach allows full-color images to be viewed during reception.

Linearity—The property found in nature and most analog electrical circuits that governs the processing and combination of signals by treating all signal levels the same way.

Load Line—A line drawn through a family of characteristic curves that shows the operating points of an analog device for a given output load impedance.

Loading—The condition that occurs when a cascaded analog stage modifies the operation of the previous stage.

Loopback—A test performed by connecting the output of a modulator to the input of a demodulator.

LORAN—Long RAnge Navigation. A hyperbolic navigation system for ships and aircraft using multiple transmitters with fixed delays between pulse transmissions. LORAN-C is a more recent version using a frequency of 100 kHz.

LOS—Line of sight.

Low-pass filter—a filter designed to pass all frequencies below a cutoff frequency, while rejecting frequencies above the cutoff frequency.

LSB—Least-significant bit.

LUF—Lowest useable frequency (LUF). This is the lowest frequency that is likely to propagate via ionospheric refraction between a particular pair of end point.

Luminance—The brightness component of a video signal. Usually computed as Y (the luminance signal) = 0.59 G (green) + 0.30 R (red) + 0.11 B (blue).

Mag-mount — antenna with a magnetic base that permits quick installation and removal from a motor vehicle or other metal surface.

Magnetron—A microwave oscillator tube developed during WWII that enabled microwave radar.

Mark—A telegraph signal element in which current is flowing.

Mastergroup—A level in the analog multiplex hierarchy encompassing five supergroups and thus 300 voice channels or equivalent.

Mean power—The average power supplied to an antenna transmission line during an interval of time sufficiently long compared with the lowest frequency encountered in the modulation taken under normal operating conditions.

Metal-Oxide Semiconductor (MOSFET) —A field-effect transistor that forms its electric field through an insulating oxide layer.

Microwave—UHF and SHF radio signals with frequencies above 1000 MHz.

Mobile Host—Also known as a "mobile node," this addressed entity in the Mobile IP protocol roams between its home network and foreign networks.

Mobile IP—This mobile industry standard enhances the IP protocol to remedy problems associated with using the standard TCP/IP with a mobile entity. It allows for transparent routing of IP data grams to mobile hosts (nodes) on the Internet.

Mode—A type of ham radio communication; examples are *frequency modulation (FM voice), slow-scan television (SSTV) and SSB (single sideband voice).*

Modem—Modulator-demodulator, a device that connects between a data terminal and communication line (or radio). Also called data set.

Monostatic—A radar system in which transmitting and receiving locations are collocated.

Morse code—A popular communication mode transmitted by on/off keying of a radio-frequency signal. Hams use the *international Morse code.*

MSB—Most-significant bit.

MSK—Frequency-shift keying where the shift in Hz is equal to half the signaling rate in bits per second.

MUF—Maximum useable frequency (MUF). This is the highest frequency that is likely to propagate via ionospheric refraction between a particular pair of end point.

Multiplex—The combining of multiple signals, such as telephone trunk connections onto a single transmission system.

NAK—Negative acknowledge (opposite of ACK).

NEC—*Numerical Electromagnetics Code.* The calculating algorithms upon which many antenna modeling programs, including *EZNEC,* are based.

Necessary bandwidth—The width of the transmitted frequency band that is just sufficient to ensure the transmission of information at the rate and with the quality required under specified conditions.

Net—An on-the-air meeting of hams at a set time, day and radio frequency.

Network—A system of interconnected radios to allow more than one station access to shared resources.

NiCd—a nickel-cadmium battery that may be recharged many times; often used to power portable transceivers. Pronounced "NYE-cad."

Node—A unique host on a network such as a printer, computer device, handheld Personal Digital Assistant (PDA), or a mainframe.

Noise—Any signal that interferes with the desired signal in electronic communications or systems.

Noise Figure (NF)—A measure of the noise added to a signal by an analog processing stage.

Nonlinear—having an output that is not in linear proportion to the input.

Notch filter—A filter that rejects or suppresses a narrow band of frequencies within a wider band of frequencies.

NRZI—Nonreturn to zero. A binary baseband code in which output transitions result from data 0s but not from 1s. Formal designation is NRZ-S (nonreturn-to-zero—space).

N-Type Impurity—A doping atom with an excess of electrons that is added to semiconductor material to give it a net negative charge.

NTSC—National Television System Committee. Television standard used in North America and Japan.

Octet—A group of eight bits.

Ohm—Unit of resistance. One ohm is defined as the resistance that will allow one ampere of current when one volt of EMF is impressed across the resistance.

Operational Amplifier (op amp)—An integrated circuit that contains a symmetrical circuit of transistors and resistors with highly improved characteristics over other forms of analog amplifiers.

Open repeater—A repeater whose access is not limited.

Orthogonal Frequency Division Multiplexing (OFDM)—A type of modulation that splits a wide frequency band into many narrow frequency bands. Both 802.11a and 802.11g use OFDM.

Originate—The station initiating a call. In modem usage, the calling station or modem tones associated therewith.

Oscillator—An unstable analog system, which causes the output signal to vary spontaneously. Also a device to produce controlled oscillations.

OSI-RM—Open Systems Interconnection Reference Model specified in ISO 7498 and CCITT Rec X.200.

Out-of-band emission (splatter)—An emission on a frequency immediately outside the necessary bandwidth caused by overmodulation on peaks (excluding spurious emissions).

Output frequency—the frequency of the repeater's transmitter (and your transceiver's receiver).

Over—a word used to indicate the end of a voice transmission.

Packet radio—A computer-to-computer radio communication mode in which information is broken into short bursts. The bursts (packets) contain addressing and error-detection information.

PAL—Phase alteration line. Television standard used in Germany and many other parts of Europe.

Parity check—Addition of non-information bits to data, making the number of ones in a group of bits always either even or odd.

Passband—the band of frequencies that a filter conducts with essentially no attenuation.

PCM—Pulse code modulation. A system in which analog voice is sampled and digitized for transmission as a series of binary coded data words.

Peak (voltage or current)—The maximum value relative to zero that an ac voltage or current attains during any cycle.

Peak Inverse Voltage (PIV)—The highest voltage that can be tolerated by a reverse biased PN junction before current is conducted.

Pentode—A five-element vacuum tube with a cathode, a control grid, a screen grid, a suppressor grid, and a plate.

PEP (peak envelope power)—The average power supplied to the antenna transmission line by a transmitter during one RF cycle at the crest of the modulation envelope taken under normal operating conditions.

Period (T)—The duration of one ac voltage or current cycle, measured in seconds (s).

Permeability (μ)—The ratio of the magnetic flux density of an iron, ferrite, or similar core in an electromagnet compared to the magnetic flux density of an air core, when the current through the electromagnet is held constant.

Phased Array Radar—A radar system in which an array of antenna elements is combined with dynamic phase shift elements to synthesize an electronically steered antenna pattern.

Phone—Emissions carrying speech or other sound information having designators with A, C, D, F, G, H, J or R as the first symbol; 1, 2 or 3 as the second symbol; E as the third symbol. Also speech emissions having B as the first symbol; 7, 8 or 9 as the second symbol; E as the third symbol.

PID—Protocol identifier. Used in AX.25 to specify the network-layer protocol used.

Pixel—Picture element. The dots that make up images on a computer's monitor.

Plate—See anode, usually used with vacuum tubes.

PN Junction—The region that occurs when P-type semiconductor material is placed in contact with N-type semiconductor material.

POS (Point of Service)—A generation of narrowband digital, two-way, low-powered wireless services in the 800-900 MHz bands that will support confirmed delivery of message, full two-way data transfer, voice messaging and connectivity via the Internet.

Power—Power is the rate at which work is done. One watt of power is equal to one volt of EMF, causing a current of one ampere through a resistor. Power is expressed in three ways: (1) Peak envelope power (PEP); (2) Mean power; and (3) Carrier power.

Power Conditioner—Another term for a power supply.

PPI—Plan position indicator. A radar indicator showing azimuth, range and target return amplitude.

PPP (Point-to-Point Protocol)—A method of connecting a computer to the Internet. PPP is more stable than the older SLIP protocol and provides error-checking features.

Primary—The master station in a master slave relationship; the master maintains control and is able to perform actions that the slave cannot. (Compare secondary.)

Protocol—A formal set of rules and procedures for the exchange of information within a network.

PSK—Phase-shift keying.

PTT—Push-to-talk. A transmit-receive switching system in which a button or switch, typically located on a microphone, is used to initiate transmission.

P-Type Impurity—A doping atom with an excess of holes that is added to semiconductor material to give it a net positive charge.

Public service—Activities involving Amateur Radio that hams perform to benefit their communities.

Q (quality factor)—The ratio of energy stored in a reactive component (capacitor or inductor) to the energy dissipated, equal to the reactance divided by the resistance.

QAM—Quadrature Amplitude Modulation. A method of simultaneous phase and amplitude modulation. The number which precedes it, e.g., 64QAM, indicates the number of discrete stages in each pulse.

QRP—An abbreviation for low power.

QSL bureau—A system for sending *QSL cards* to and from ham radio operators.

QSL cards—Cards that serve to confirm communication between two hams.

QST—The premiere ham radio monthly magazine, published by the *ARRL*. *QST* means "calling all radio amateurs."

RACES (Radio Amateur Civil Emergency Service)—A radio service that uses amateur stations for civil defense communication during periods of local, regional or national civil emergencies.

Radiated emission—radio-frequency energy that is coupled between two systems by electromagnetic fields.

Radio (or Ham) shack—The room where Amateur Radio operators keep their station.

Radio-frequency interference (RFI)—interference caused by a source of radio-frequency signals. This is a subclass of EMI.

Radio Regulations—The latest ITU *Radio Regulations*.

Radiotelegraphy—See *Morse code*.

RAM—Random access memory.

Reactance (X)—Opposition to alternating current by storage in an electrical field (by a capacitor) or in a magnetic field (by an inductor), measured in ohms (Ω).

Receiver—A device that converts radio signals into a form that can be heard.

Regulator—A device (such as a Zener diode) or circuitry in a power supply for maintaining a constant output voltage over a range of load currents and input voltages.

Remote control—The use of a control operator who indirectly manipulates the operating adjustments in the station through a control link to achieve compliance with the FCC Rules.

Remote Presence—The ability to establish remote network connections and still appear to be connected to the home network.

Repeater—An amateur station, usually located on a mountaintop, hilltop or tall building, that receives and simultaneously retransmits the signals of other stations on a different channel or channels for greater range.

Repeater Directory—an annual ARRL publication that lists repeaters in the US, Canada and other areas.

Resistance (R)—Opposition to current by conversion into other forms of energy, such as heat, measured in ohms (Ω).

Resonance—Ordinarily, the condition in an ac circuit containing both capacitive and reactance, where maximum RF current flows.

RF (Radio frequencies)—The range of frequencies that can travel through space in the form of electromagnetic radiation.

RGB—Red, Green, Blue. One of the models used to represent colors. Due to the characteristics of the human eye, most colors can be simulated by various blends of red, green, and blue light.

RIC (Radio interface card)—A PCMCIA device with an antenna port used in *HSMM* radio to allow a personal computer to control a radio transceiver.

Ripple—The residual ac left after rectification, filtration and regulation of the input power.

RMS (voltage or current)—Literally, "root mean square," the square root of the average of the squares of the instantaneous values for one cycle of a waveform. A dc voltage or current that will produce the same heating effect as the waveform. For a sine wave, the RMS value is equal to 0.707 times the peak value of ac voltage or current.

Robot—(1) Abbreviation for Robot 1200C scan converter; (2) a family of SSTV transmission modes introduced with the 1200C.

Router—A network packet switch. In packet radio, a network-level relay station capable of routing packets.

RS-232-C—See EIA-232-C.

RTTY—Narrow-band direct-printing telegraphy emissions having designators with A, C, D, F, G, H, J or R as the first symbol; 1 as the second symbol; B as the third symbol; and emission J2B.

RTS—Request to send, physical-level signal used to control the direction of data transmission of the local DCE.

RTTY—Radioteletype.

RxD—Received data, physical-level signals generated by the DCE are sent to the DTE on this circuit.

SAREX (Space Amateur Radio Experiment)—Amateur Radio equipment flown in space and operated by astronauts who are licensed Amateur Radio operators.

Saturation Region—The region in the characteristic curve of an analog device in which the output signal can be made no larger. See *Clamping*.

Scan converter—A device that converts one TV standard to another. For example, the Robot 1200C converts SSTV to and from FSTV.

Scottie—A family of amateur SSTV transmission modes developed by Eddie Murphy, GM3SBC, in Scotland.

Sea clutter—The undesired reflection from ocean waves that mask low altitude radar targets.

SECAM—Sequential color and memory. Television standard used in France and the Commonwealth of Independent States.

Secondary—The slave in a master-slave relationship. Compare primary.

Secondary Breakdown—A runaway failure condition in a transistor, occurring at higher collector-emitter voltages, where hot spots occur due to (and promoting) localization of the collector current at that region of the chip.

Security—The ability to create secure channels for user authentication, data integrity, and data privacy.

Semiconductor—An elemental material whose current conductance can be controlled.

Separation or **split**—the difference (in kHz) between a repeater's transmitter and receiver frequencies. Repeaters that use unusual separations, such as 1 MHz on 2 m, are sometimes said to have "oddball splits."

Series Pass Transistor, or Pass Transistor—The transistor(s) that controls the passage of power between the unregulated dc source and the load in a regulator. In a linear regulator, the series pass transistor acts as a controlled resistor to drop the voltage to that needed by the load. In a switch-mode regulator, the series pass transistor switches between its ON and OFF states.

Service Set Identity (SSID)—The identification for an AP. It is transmitted continuously in the form of a beacon.

Signal-To-Noise Ratio (SNR)—The ratio of the strength of the desired signal to that of the unwanted signal (noise).

Simplex—a mode of communication in which users transmit and receive on the same frequency.

Slew Rate—The maximum rate at which a signal may change levels and still be accurately amplified in a particular device.

SOAR (Safe Operating ARea)—The range of permissible collector current and collector-emitter voltage combinations where a transistor may be safely operated without danger of device failure.

Source—The connection at one end of the channel of a field-effect transistor, often the reference. In packet radio, the station transmitting the frame over a direct radio link or via a repeater.

Space—A telegraph signal element in which current is not flowing.

Space station—An amateur station located more than 50 km above the Earth's surface.

Spike—An extremely short perturbation on a power line, usually lasting less than a few microseconds.

Splatter—See *Out-of-band emission*.

Spread Spectrum—A technology, originated during World War II, which distributes or spreads a radio signal over a broad frequency range. This spreading prevents narrow band signals and noise sources from interfering with the spread spectrum signal. The spread spectrum signal is heard as noise to the traditional narrow band receiver.

Spurious emission—An emission, on frequencies outside the necessary bandwidth of a transmission, the level of which may be reduced without affecting the information being transmitted. They include harmonic emissions, intermodulation products and frequency conversion products, but exclude out-of-band emissions.

SSB (Single sideband)—A common *mode* of voice operation on the amateur bands.

SSID—Secondary station identifier. In AX.25 link-layer protocol, a multipurpose octet to identify several packet radio stations operating under the same call sign.

SSTV (Slow-scan television)—A *mode* of operation in which ham radio operators exchange still pictures from their stations.

Superhet—Superheterodyne. A radio receiver architecture in which signals are mixed or heterodyned to a different frequency for processing.

Supergroup—A combination of five multiplexed voice channel groups into a single higher bandwidth baseband system carrying 60 voice channels or equivalent.

Supermastergroup—A combination of three multiplexed voice channel supergroups into a single higher bandwidth baseband analog system carrying 900 voice channels or equivalent.

Superposition—The natural process of adding two or more signals together and having each signal retain its unique identity.

Surge—A moderate-duration perturbation on a power line, usually lasting for hundreds of milliseconds to several seconds.

Susceptance (B)—The reciprocal of reactance, measured in siemens (S).

Susceptibility—the characteristic of electronic equipment that permits undesired responses when subjected to electromagnetic energy.

SWL (Shortwave listener)—A person who enjoys listening to shortwave radio broadcasts or Amateur Radio conversations.

SWR—Standing wave ratio. The ratio of maximum to minimum voltage on a transmission line. A measure of how well it is matched to the impedance of the load.

Sync—That part of a TV signal that indicates the beginning of a frame (vertical sync) or the beginning of a scan line (horizontal sync).

TAPR—Tucson Amateur Packet Radio Corporation, a nonprofit organization involved in packet-radio development.

TDM (Time Division Multiplex)—A system that combines multiple voice or data streams into a single time sequenced signal for transmission.

TDMA (Time Division Multiple Access)—A digital radio system that separates users by time.

Telecommand—A one-way transmission to initiate, modify, or terminate functions of a device at a distance.

Telecommand station—An amateur station that transmits communication to initiate, modify, or terminate functions of a space station.

Telemetry—A one-way transmission of measurements at a distance from the measuring instrument.

Teleport—A radio station that acts as a relay between terrestrial radio stations and a communications satellite.

Test—Emissions containing no information having the designators with N as the third symbol. Test does not include pulse emissions with no information or modulation unless pulse emissions are also authorized in the frequency band.

Tetrode—A four-element vacuum tube with a cathode, a control grid, a screen grid, and a plate.

Third Party Mobile IP—An Internet technology solution that provides both wireless and wire line IP network and media roaming/communications to both Intranet and Internet services.

Throughput—The amount of data processed, or transferred from one place to another in a specified amount of time. Data transfer rates for disk drives and networks are measured in terms of throughput. Typically, throughput is measured in kbit/s, Mbit/s, and Gbit/s.

TIA (Telecommunications Industry Association)—Telecommunications Industry Association, 2500 Wilson Blvd, Arlington, VA 22001. On the web: **www.tiaonline.org**.

Time constant—The time required for the voltage in an RC circuit or the current in an RL circuit to rise from zero to approximately 63.2% of its maximum value or to fall from its maximum value 63.2% toward zero.

Timer—a device that measures the length of each transmission and causes the repeater or a repeater function to turn off after a transmission has exceeded a certain length.

TIS (Technical Information Service)—A service of the *ARRL* that helps hams solve technical problems.

TNC—Terminal node controller, a device that assembles and disassembles packets (frames); sometimes called a PAD.

Tone Pad—an array of 12 or 16 numbered keys that generate the standard telephone dual-tone multifrequency keypad. (see *autopatch*).

(*DTMF*) dialing signals. Resembles a standard telephone

Toroid—Literally, any donut-shaped solid; most commonly referring to ferrite or powdered-iron cores supporting inductors and transformers.

Transceiver—A radio transmitter and receiver combined in one unit.

Transient—A short perturbation on a power line, usually lasting for microseconds to tens of milliseconds.

Transducer—Any device that converts one form of energy to another; for example an antenna, which converts electrical energy to electromagnetic energy or a speaker, which converts electrical energy to sonic energy.

Transformer—A device consisting of at least two coupled inductors capable of transferring energy through mutual inductance.

Transmitter—A device that produces radio-frequency signals.

Transponder—A system that transmits a signal in response to the appropriate received signal.

Triode—A three-element vacuum tube with a cathode, a grid, and a plate.

TRF—A tuned radio frequency receiver. This technology used multiple tuned RF stages at the signal frequency ahead of a detector to provide gain and selectivity. This architecture preceded the superheterodyne receiver.

TR switch—Transmit-receive switch to allow automatic selection between receive and transmitter for one antenna.

TTY—Teletypewriter.

Turnaround time—The time required to reverse the direction of a half-duplex circuit, required by propagation, modem reversal and transmit-receive switching time of transceiver.

TVI—interference to television systems.

TxD—Transmitted data, physical-level data signals transferred on a circuit from the DTE to the DCE.

UHF (Ultra-high frequencies)—The radio frequencies from 300 to 3000 MHz.

UI—Unnumbered information frame.

Unipolar Transistor—see *Field-Effect Transistor (FET)*.

V.24—A CCITT standard defining physical-level interface circuits between a DTE (terminal) and DCE (modem), equivalent to EIA RS-232-C.

V.28—A CCITT standard defining electrical characteristics for V.24 interface.

Varistor—A surge suppression device used to absorb transients and spikes occurring on the power lines, thereby protecting electronic equipment plugged into that line. Frequently, the term MOV (*Metal Oxide Varistor*) is used instead.

VCO—Voltage controlled oscillator. An oscillator in which the frequency can be varied by adjustment of the voltage applied to a control element in the oscillator circuit.

VE (Volunteer Examiners)—Amateur Radio operators who give Amateur Radio licensing examinations.

VHF (Very-high frequencies)—The radio frequencies from 30 to 300 MHz.

Virtual circuit—A mode of packet networking in which a logical connection that emulates a point-to-point circuit is established (compare Datagram).

VIS—Vertical Interval Signaling. Digital encoding of the transmission mode in the vertical sync portion of an SSTV image. This allows the receiver of a picture to automatically select the proper mode. This was introduced as part of the Robot modes and is now used by all SSTV software designers.

Volt—A measure of electromotive force.

Voltage (E)—Electromotive force or electrical pressure, measured in volts (V).

Volt-Amperes—The product obtained by multiplying the current times the voltage in an ac circuit without regard for the phase angle between the two. This is also known as the *apparent power* delivered to the load as opposed to the actual or *real power* absorbed by the load, expressed in watts.

Voltage Multiplier—A type of rectifier circuit that is arranged so as to charge a capacitor or capacitors on one half-cycle of the ac input voltage waveform, and then to connect these capacitors in series with the rectified line or other charged capacitors on the alternate half-cycle. The voltage doubler and tripler are commonly used forms of the voltage multiplier.

VOR—VHF omnidirection range. A navigation system designed for aircraft use. Appropriate receivers provide bearings to VOR locations, located at major airports.

WAS (Worked All States)—An *ARRL* award that is earned when an Amateur Radio operator talks to and exchanges QSL cards with a ham in each of the 50 states in the US.

WAVE (Worked All VE)—An award that is earned when a ham talks to and exchanges QSL cards with a ham in each Canadian province.

Wavelength—The distance a readio wave travels during one cycle. A means of designating a frequency *band*, such as the 80-meter band.

Window—In packet radio at the link layer, the range of frame numbers within the control field used to set the maximum number of frames that the sender may transmit before it receives an acknowledgment from the receiver.

Windows OS—Microsoft *Windows* Operating System.

Wireless Data—Information or "intelligence," sent or received by wireless transmission/reception without the direct aid of a landline.

Wired Equivalent Privacy (WEP)—A standard for providing minimal privacy of wireless LAN communication by encrypting individual data frames.

Wireless Fidelity (Wi-Fi)—The Wireless Ethernet Compatibility Alliance certification program to ensure that equipment claiming to be in compliance with 802.11 standards is truly interoperable. The term Wi-Fi5 is sometimes applied to 802.11a equipment that operates on the 5-GHz band.

WLAN (Wireless Local Area Network)—A local-area network that uses high frequency radio waves rather than wires to communicate between nodes.

Work—To contact another ham.

Wraase—A family of amateur SSTV transmission modes first introduced with the Wraase SC-1 scan converter developed by Volker Wraase, DL2RZ, of Wraase Electronik, Germany.

X.25—CCITT packet-switching protocol.

Zener Diode—A PN-junction diode with a controlled peak inverse voltage so that it will start conducting current at a preset reverse voltage.

Index

FEEDBACK

Please use this form to give us your comments on this book and what you'd like to see in future editions, or e-mail us at **pubsfdbk@arrl.org** (publications feedback). If you use e-mail, please include your name, call, e-mail address and the book title, edition and printing in the body of your message. Also indicate whether or not you are an ARRL member.

Where did you purchase this book?
□ From ARRL directly   □ From an ARRL dealer

Is there a dealer who carries ARRL publications within:
□ 5 miles   □ 15 miles   □ 30 miles of your location?   □ Not sure.

License class:
□ Novice   □ Technician   □ Technician with code   □ General   □ Advanced   □ Amateur Extra

Name _____

ARRL member? □ Yes □ No

Call Sign _____

Daytime Phone (   ) _____

Age _____

Address _____

City, State/Province, ZIP/Postal Code _____

If licensed, how long? _____

e-mail address: _____

Other hobbies _____

_____

Occupation _____

For ARRL use only	BR
Edition	1 2 3 4 5 6 7 8 9 10 11 12
Printing	2 3 4 5 6 7 8 9 10 11 12

From _____

_____

_____

EDITOR, BASIC RADIO
ARRL—THE NATIONAL ASSOCIATION FOR AMATEUR RADIO
225 MAIN STREET
NEWINGTON CT 06111-1494

— — — — — — — — — — — — — — — — please fold and tape — — — — — — — — — — — — — — — — — —